Phase transitions

Introduction with worked examples

Harshad Bhadeshia and Haixue Yan

i

Phase Transitions

An introduction with worked examples

Harshad K. D. H. Bhadeshia

Haixue Yan

Queen Mary University of London
School of Engineering and Materials Science

PHASE TRANSITIONS – an introduction with worked examples

Arising from lectures given at Queen Mary University of London, this book presents an introductory theme of "Phase transitions". Also included are worked examples that emphasise understanding.

Phase transitions – an introduction with worked examples features contributions by Harshad K. D. H. Bhadeshia and Haixue Yan.

The subjects covered range from mechanisms, thermodynamics, kinetics and applications.

PUBLISHED INDEPENDENTLY BY THE AUTHORS
Mile End Road
London E1 4NS United Kingdom

https://www.sems.qmul.ac.uk/

First published 2023

Phase transitions – an introduction with worked examples
        written by Harshad K. D. H. Bhadeshia and Haixue Yan
Includes index

ISBN 9798395358547 Paperback

The publishers bear no responsibility for the persistence or accuracy
of URLs for external or third-party internet websites referred to in this publication
and do not guarantee that any content on such websites is, or will remain,
accurate or appropriate.

Typeset in Computer Modern font
by the authors

Contents

Preface

This short book is an *introduction*. Each of the concepts covered is sufficient and complete to the extent appropriate for an early semester of an undergraduate course. There are worked examples that can stimulate absorption of the subject. The book is suitable for second year students in any of the following disciplines: materials science, engineering, chemical engineering, physics, chemistry, Earth sciences and physical sciences in general. References are provided at the end for anyone who wants, for whatever reason, to delve deeper.

People use materials routinely and without worrying about the structure within. If the material functions reliably then there is no justification to think further. There are 9 million scientists in the world today, of whom about 300,000 are concerned with materials – i.e., a fraction 0.00004 of the world's population. This very special minority drives the creation of new materials by probing and manipulating the structure of materials. Structure itself is as defined over length scales that range from atoms to engineering dimensions so the subject is inherently interdisciplinary like no other.

Countless variables can influence structure and therefore, properties. For this reason, logic is needed to understand, and express quantitatively how such parameters work. One example is that every element in the periodic table can, to a greater or lesser extent, dissolve in every other element. This represents far too much work to explore experimentally. But the mixing and its consequences can in principle be expressed using free energies and atomic mobilities to calculate what should happen. This would then be validated experimentally and used to inspire new materials and processes.

Throughout this book, braces are used to imply a functional relationship, i.e., $f\{x\}$ means that f is a function with an argument x.

We are grateful to many who were kind enough to provide us with images and information, as acknowledged appropriately in the text.

Harshad K. D. H. Bhadeshia and Haixue Yan

The year 2023
London.

Chapter 1

Introduction

Ideally, a *phase* is a homogeneous and physically distinct region. It is distinguished by properties such as composition and the arrangement of atoms within, or even readily measurable properties such as electrical resistance or density. A phase transformation can be stimulated by changes in temperature, stress, magnetic fields, electrical stimuli and the pull of gravity. A *transformation* means that the pattern in which atoms are arranged may change; this may even involve a configuration of atoms that has no periodicity. A phase can be gaseous, liquid, plasma or solid. Transformations can involve entities different from atoms, for example, dipoles or magnetic moments.

The ability to control transformations plays a huge role in the design of materials for specific purposes, or simply to satisfy curiosity. Transformations in particular form the basis of structure control at a variety of length scales. We shall attempt to address why phase changes occur and the mechanism by which they occur.

1.1 MECHANISMS OF TRANSFORMATION

One of the reasons why there is a great number of microstructures at our disposal is because the atoms can move in a variety of ways to achieve the same change in crystal structure. The shape and packing of the resultant crystals will depend on this mechanism. The transformation can occur either by breaking all the bonds and rearranging the atoms into an alternative pattern (*reconstructive* transformation), or by homogeneously deforming the original pattern into a new crystal structure, i.e. *displacive* transformation, Figure 1.1.

In the displacive mechanism the change in crystal structure also alters the macroscopic shape of the sample when the latter is not constrained. The shape deformation during constrained transformation is accommodated by a combination of elastic and plastic strains in the surrounding matrix. The product phase then grows in the form of thin plates to minimise the strains. The atoms are displaced into their new positions in a coordinated motion. Displacive transformations, therefore, occur at temperatures where diffusion is not possible during the time scale of the experiment. Some solutes may be forced into the product phase, even if under equilibrium they would prefer not to be there. Both the trapping of atoms and the strains make displacive transformations less favourable from a thermodynamic point of view. However, if atoms lack mobility, e.g., at low temperatures,

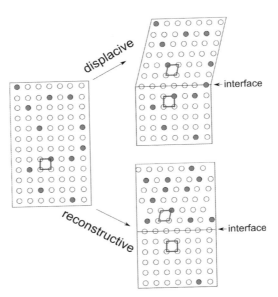

FIGURE 1.1 Schematic illustration of the mechanisms of transformation. The parent crystal on the left contains two kinds of atoms; the filled circles are solute atoms given their small concentration relative to the open circles that identify solvent atoms. The figures on the right represent partially transformed samples with the parent and product unit cells outlined in bold. The displacive transformation is accompanied by a large change in shape that is consistent with the change in atomic pattern, but the chemical composition remains unaltered. Whereas there is no change in shape during the reconstructive transformation illustrated, the different atoms have partitioned between the two phases depending on the equilibrium solubilities within those phases.

this is the only mechanism available to achieve the transformation.

It is the ability of atoms to diffuse that leads to the new crystal structure during a reconstructive transformation. Imagine that the transformation proceeds as by the displacive mechanism (Figure 1.2a,b), but that the resulting shape deformation is eliminated by transporting the segment as in Figure 1.2c to recover the overall shape shown in (d). This transport is the diffusion that is needed in order that the strain energy term is essentially eliminated, as if there is fluid flow in the surrounding matrix. The flow of matter is sufficient to avoid any shear components of the shape deformation, leaving only the effects of volume change. In alloys, the diffusion process may also lead to the redistribution of solutes between the phases in a manner consistent with a reduction in the overall free energy.

Figure 1.3a shows an atomic force microscope image taken from a steel sample that was polished organically flat before it underwent a displacive transformation from a face-centred cubic to a body-centred cubic crystal structure. The sample changes its shape in the region where transformation happened, leading to sub-

2

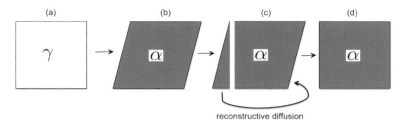

FIGURE 1.2 Phenomenological interpretation of reconstructive transformation. The virtual operation that eliminates the shape change is the diffusion required during a reconstructive transformation, irrespective of whether it occurs in a pure substance or in an alloy.

stantial upheavals of the surface, caused by the coordinated motion of iron atoms without any diffusion. The upheavals can be characterised to consist of a shear strain of about 0.26. There is also a small volume change (0.03) by dilatation normal to the plane on which the shear occurs.

In contrast, Figure 1.3b shows a tomographic image of the profuse precipitation of Ni_4Ti_3 in a solid solution of nickel and titanium; the precipitates have quite a different chemical composition than the matrix so diffusion accompanied transformation. The precipitation process is a reconstructive transformation with no surface upheavals of the type observed in Figure 1.3a.

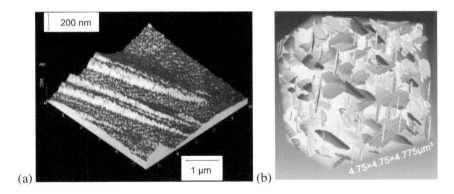

FIGURE 1.3 (a) Atomic force image showing the upheaval of the initially flat surface due to displacive transformation in steel, from face-centred cubic to body-centred cubic crystal structure. The sample has been polished flat prior to transformation [1]. (b) Three-dimensional characterisation showing precipitation of Ni_4Ti_3 in a Ni-Ti solid solution. Image courtesy of Dominique Schryvers and Cao Sanshan of the Electron Microscopy for Materials Science section of the University of Antwerp.

1.2 CONDITION FOR DISPLACIVE TRANSFORMATION

A displacive transformation occurs only when atomic mobility is limited. The velocity of the interface during displacive transformation will be greater than the ability of atoms to diffuse. This can be expressed quantitatively by comparing the rate at which the interface moves with the corresponding *diffusion velocity*:

$$\underbrace{v}_{\text{interface velocity}} > \underbrace{\frac{D}{\lambda}}_{\text{diffusion velocity}} \qquad (1.1)$$

where λ is the distance between adjacent atomic positions. Notice that a diffusion coefficient has units of $m^2\,s^{-1}$ so dividing by λ gives a speed with units $m\,s^{-1}$, which is the fastest speed with which the atoms can diffuse since λ is the smallest distance that an atom can jump.

Example 1: barium titanate

The ceramic barium titanate has a variety of crystal structures, two of which are illustrated in Figure 1.4. By working out the number of atoms of each kind in the unit cell of the cubic form, determine the chemical formula of the oxide. Bear in mind that atoms located at corners or faces of the unit cell will be shared with other unit cells, i.e., they cannot be assigned completely to the unit cell drawn.

In the tetragonal cell, the x and y axes are slightly smaller than in the case of the cubic lattice, and the z-axis (vertical) is slightly elongated. Note particularly that the titanium atom does not lie at the centre of the cell but is slightly elevated along z. Because of this asymmetry, the centre of mass of the ions will move when the crystal is deformed. Is there any consequence of this?

Would the transformation of the cubic to slightly tetragonal lattice during cooling at 120 °C be displacive or reconstructive?

Solution 1

Each atom at a corner of the unit cell contributes just $\frac{1}{8}$ to the cell; any at the face-centres a $\frac{1}{2}$; the titanium atom is wholly enclosed by the cell, so the chemical formula is $BaTiO_3$.

If the crystal is deformed homogeneously, the centre of mass shifts, leading to the development of a voltage, i.e., piezoelectricity.

The transformation is displacive, first because the transformation temperature is

4

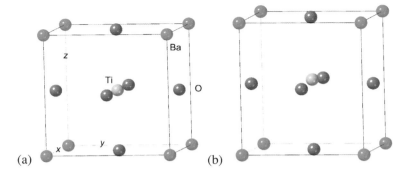

FIGURE 1.4 Examples of the crystal structures of barium titanate. (a) Cubic lattice. (b) Tetragonal lattice, with the titanium atom not located at the centre of the cell, but at elevated coordinates $(\frac{1}{2}, \frac{1}{2}, 0.52)$ along the z axis. Similarly, the oxygen atoms on the vertical faces have coordinates equivalent to $\frac{1}{2}, 1, 0.52)$ and the oxygen atom on the horizontal faces are at coordinates equivalent to $(\frac{1}{2}, \frac{1}{2}, 0.96)$.

too low for diffusion to be possible during the time scale of the experiment – a transition like this can occur in a microsecond or so [2]. The second reason is that the distortion needed to change the cubic to tetragonal lattice is really quite small. The lattice parameter of the cubic form close to the Curie point is 0.4009 nm whereas the tetragonal form at the same temperature has $c = 0.4022$ nm and $a = 0.4003$ nm [3]. The distortions are therefore slight. The volume of the tetragonal cell is smaller than the cubic titanate at the same temperature.

In fact, there are two further even lower temperature transformations that occur in barium titanate: at $0\,°C$ it changes from tetragonal to orthorhombic, and then below $-90\,°C$ from an orthorhombic to a rhombohedral crystal structure. All of these are spontaneously polarised, along the $\langle 001 \rangle$, $\langle 110 \rangle$ and $\langle 111 \rangle$ along the tetragonal, orthorhombic and rhombohedral directions, respectively.

Apart from its piezoelectric properties, the tetragonal $BaTiO_3$ has a permanent dipole that can be switched in an applied electric field, which also makes it ferroelectric. In this state, the dipoles within the crystal can interact and become aligned in the same direction over a part of the crystal. In other parts, the dipoles can align on different $\langle 100 \rangle$ directions (different domains) so that in the absence of a field, the material does not exhibit a net dipole moment. The application of an electrical field favours the growth of those domains that best align to the field so a net polarisation results.

Barium titanate also has a large *relative permittivity*. This means that when placed in an electrical field, the magnitude of that field is greatly reduced within the titanate. The relative permittivity is dimensionless and frequency dependent;

it is the ratio of the capacitance of a capacitor consisting of two conducting plates separated by the barium titanate, to the capacitance when there is a vacuum between those plates. The relative permittivity of barium titanate is found to be in the range 7000-15000 at 1 kHz, which is much greater than common ceramics or polymers that have values < 100. This means that it is an insulator, and hence is used to make capacitors for electronic applications.

1.3 ELEMENTARY PRINCIPLES OF X-RAY DIFFRACTION

One of the characterisation methods used in the study of phase transformations is X-ray diffraction. The method can help in the determination of crystal structures, such as the cubic and tetragonal forms of barium titanate, the fractions of phases present in a mixture of phases, and even the positions of atoms within the unit cell. In a conventional laboratory, the penetration of X-rays into a metal is a few micrometres; using synchrotron X-rays provided for academic studies by a variety of national and international organisations, permits the use of far more intense X-rays, capable of penetrating 10 mm of a metallic sample and allowing a time-resolved study. Synchrotron facilities cost billions to construct, but can be used free if a good scientific case is made and if the information obtained is made openly available. However, here we deal with a simple case to illustrate the diffraction.

Consider waves of length λ incident on planes of atoms. The beams reflected from *different* planes in the parallel set illustrated in Fig. 1.5 must be in phase to avoid destructive interference. The path difference between beams a and b, i.e. the distance xyz, must then be an integral number of wavelengths. Since $xyz = 2d \sin \theta$, the diffraction condition is

$$n\lambda = 2d \sin \theta \tag{1.2}$$

where n is an integer; this equation is the Bragg law [4], with θ designated the Bragg angle.

Therefore, when a crystal is irradiated with X-rays, the intensity of diffracted waves will only be significant at particular values of θ, from which it is possible to deduce some information about the crystal structure of the sample.

6

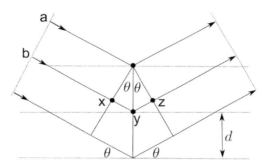

FIGURE 1.5   Electromagnetic waves incident on a set of parallel crystal planes with an interplanar spacing d. The angle of emergence of the scattered waves is the same as that of incidence.

Example 2: X-rays from cubic & tetragonal barium titanate

(i) A crystal with a hexagonal unit cell has the parameters $a = 1$ nm and $c = 2$ nm. Select the correct d-spacing of the $\{001\}$ ($\equiv \{0001\}$ in four index notation) planes: (a) 1.52 nm; (b) 1.71 nm; (c) 2 nm.

(ii) We have seen that barium titanate can exist both in the cubic and tetragonal crystal structures. Assuming that the lattice parameters are $a_{\text{cubic}} = 0.404$ nm, and $a_{\text{tetragonal}} = b_{\text{tetragonal}} = 3.9$ nm with $c_{\text{tetragonal}} = 4.1$ nm, use an X-ray wavelength of 0.15405 nm to calculate the positions 2θ of diffraction peaks resulting from $\{100\}$, $\{110\}$, $\{111\}$ planes.

Explain why the $\{100\}$, (001) peaks separate in the case of the tetragonal lattice, but the $\{111\}$ does not. Note that the level of tetragonality has been exaggerated for clarity.

(iii) Barium titanate undergoes a cubic to tetragonal transformation on cooling below the Curie point of 120 °C. In the tetragonal cell, the titanium atom no longer lies at the centre of the cell, but is displaced along either $[001]_{\text{tetragonal}}$ or $[00\bar{1}]_{\text{tetragonal}}$. Suppose a large single crystal of cubic titanate transforms on cooling below the Curie point, giving your reasoning, explain how many different orientations of the tetragonal cell could in principle be observed?

Solution 2

(i) 2 nm since we are dealing with basal planes.

(ii) Bearing in mind that the d-spacing for a plane with Miller indices $\{hk\ell\}$ is, for an orthogonal cell, given by the equation

$$\frac{1}{d^2_{\{hk\ell\}}} = \frac{h^2}{a^2} + \frac{k^2}{b^2} + \frac{\ell^2}{c^2}, \tag{1.3}$$

and using the Bragg Equation 1.2, the positions of the peaks on Figure 1.6 can be calculated; a, b and c are the lattice parameters.

For the tetragonal lattice the $\{100\}$, (001) planes must have different spacings because $c > a = b$. Therefore, the $\{100\}_{\text{cubic}}$ peak splits into two. Notice that $\{100\}_{\text{tetragonal}}$ is more intense than $(001)_{\text{tetragonal}}$ because $\{100\}_{\text{tetragonal}}$ contains intensity from both $(100)_{\text{tetragonal}}$ and $(010)_{\text{tetragonal}}$ whereas there is only one set of $(001)_{\text{tetragonal}}$ planes normal to the c axis. The $\{111\}_{\text{tetragonal}}$ peak does not split because all such planes i.e.,

$\left[(111)_{\text{tetragonal}}, (11\bar{1})_{\text{tetragonal}}, (1\bar{1}1)_{\text{tetragonal}}, (\bar{1}11)_{\text{tetragonal}}\right]$ all have indices that include the c-axis.

(iii) Six orientations, accounting for all the axes of the cubic cell, and their opposites.

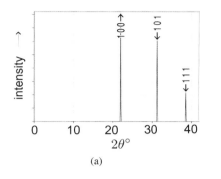
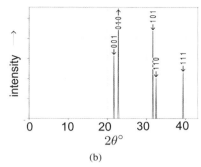

FIGURE 1.6   Barium titanate. (a) Cubic. (b) Tetragonal

Example 3: mechanism of transformation

(i)   Comment on each of these statements, giving reasons why they may or may not be be correct.

- The strain energy accompanying a displacive transformation is much greater than that associated with a reconstructive transformation.
- A displacive transformation product is as expected from equilibrium considerations.
- Adequate diffusion is a necessary condition for displacive transformations.

(ii)   After elastic deformation, a material exhibits an ideal plastic behaviour (Figure 1.7a). Assuming that all the energy of both elastic and plastic deformation is stored within the material, calculate the resulting stored energy per unit volume. Show each step in your calculation.

(iii) Why do displacive transformations occur even though they lead to microstructures that are far from equilibrium? The diffusion coefficient in a particular solid where the average interatomic spacing is 0.2 nm is given by $D = 10^{-14}\,\mathrm{m^2\,s^{-1}}$. A phase transformation product in this system, is observed to grow at $2 \times 10^{-7}\,\mathrm{m\,s^{-1}}$ at 600 K. Giving reasons, explain whether this product is likely to grow by a displacive or reconstructive transformation mechanism?

(iv)   What is the property change expected when cubic barium titanate undergoes a displacive transformation into a tetragonal form? Figure 1.7b shows the X-ray diffraction pattern from the tetragonal BaTiO₃. Sketch how the labelled peaks are expected to change when it transforms into the

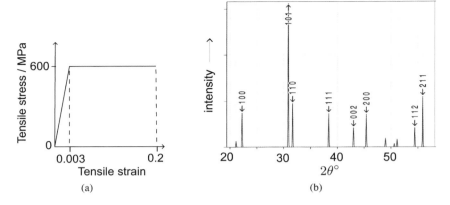

FIGURE 1.7 (a) Stress-strain curve. (b) An X-ray diffraction pattern from tetragonal $BaTiO_3$.

cubic form and justify your sketch.

(v) Which mechanism of solid-state phase transformation is likely to result in a finer structure (smaller individual crystals), displacive or reconstructive? Give two reasons to justify your answer.

Solution 3

(i) The first statement is correct because a displacive transformation leads to a change in shape that is much greater than that associated with reconstructive changes. This is because during a displacive transformation, the pattern in which the atoms are arranged is forced to change into a new one without diffusion. The second is wrong because not only is there a strain energy due to the change in shape but the chemical composition does not change even though some atoms would prefer to partition between the phases. The third is incorrect, diffusion is entirely unnecessary for a displacive transformation.

(ii) The stored energy per unit volume is simply the total area under the curve, $0.5 \times 0.003 \times 600 \times 10^6 + (0.2 - 0.003) \times 600 \times 10^6 = 119 \times 10^6 \, \mathrm{J\,m^{-3}}$. This example illustrates the fact that a deformation occurring inside a material due to a phase transformation will lead to strain energy. The product phase can then adopt a shape that minimises that strain energy, for example a plate shape.

(iii) They occur when the parent lattice is supercooled to such a temperature that atomic mobility is limited. Given atomic mobility, they would not occur. This is expressed formally using equation 1.1. The diffusion velocity, i.e. the ability of atoms to jump out of the path of the interface, is

$$\frac{D}{\lambda} = \frac{10^{-14}}{0.2 \times 10^{-9}} = 5 \times 10^{-5} \, \mathrm{m \, s^{-1}},$$

which is much greater than the given growth-velocity of $2 \times 10^{-7} \, \mathrm{m \, s^{-1}}$, so the transformation mechanism is likely to be reconstructive.

(iv) Six orientations, accounting for all the axes of the cubic cell, and their opposites.

All peaks in which the third index is non-zero split when the lattice is tetragonal with $c > a$. Therefore, in the cubic form, the [101] and [110] peaks would combine, as would [002] and [200], and [112] and [211].

(v) A displacive transformation would lead to a finer structure because the strain energy associated with the transformation ensures thin plates. Students will not know this, but the effective grain size of a thin plate is just twice the thickness. So just by phase transformation, it is possible to obtain grains that have an effective grain size of just 40 nm [5, 6].

Secondly, displacive transformations occur only at temperatures where the atomic mobility is limited. Therefore, they cannot, for example, grow across the grain boundaries of the parent phase.

Thirdly, displacive transformations occur at large driving forces (greater undercooling), therefore, the nucleation rate is greater than reconstructive transformations.

Chapter 2

Elementary thermodynamics

2.1 INTRODUCTION

Thermodynamics facilitates the linking of many observable properties so they can be seen to be a consequence of a few. It provides a firm basis for the rules that *macroscopic* systems follow at equilibrium. Nothing more needs to be said to justify its existence as a subject.

2.2 DEFINITIONS

2.2.1 Internal energy and enthalpy

The change in the internal energy ΔU of a closed system can be written as

$$\Delta U = q - w \qquad (2.1)$$

where q is the heat transferred into the system and w, the work done by the system. The sign convention is that heat added and work done by the system are positive, whereas heat given off and work done on the system are negative. Equation 2.1 may be written in differential form as

$$dU = dq - dw. \qquad (2.2)$$

For the special case where the system does work against a constant atmospheric pressure, this becomes

$$dU = dq - PdV \qquad (2.3)$$

where P is the pressure and V the volume.

The specific heat capacity of a material represents its ability to absorb or emit heat during a unit change in temperature. Heat changes the distribution of energy amongst the particles in the system (atoms, electrons, ...) and it is these fundamental mechanisms that control the heat capacity, defined formally as dq/dT. Since $dq = dU + PdV$, the specific heat capacity measured at constant volume is given by:

$$C_V = \left(\frac{\partial U}{\partial T} \right)_V.$$

A major contribution to the heat capacity of a solid comes from lattice vibrations. Each atom can be considered to be a simple harmonic oscillator with three degrees of freedom (three orthogonal directions along which it can vibrate). Each such oscillator will therefore have an average kinetic energy $3 \times \frac{1}{2}kT$ and average potential energy $3 \times \frac{1}{2}kT$ giving a net of $3kT$. Suppose there are N atoms in the solid then the energy is $3NkT$ and the contribution of lattice vibrations to heat capacity is therefore $C_V^L = 3Nk$. This supposes, however, that the vibrations of each atom are independent, whereas they assume more collective modes as the temperature is reduced. There are various methods of taking this into account, but the essence of the behaviour is approximated by treating this as an elastic-continuum problem, in Figure 2.1. It is only beyond the Debye temperature that the heat capacity becomes $3Nk$. The Debye temperatures of copper, iron and lead are, respectively, 343, 470 and 105 K.

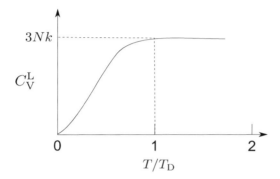

FIGURE 2.1 The Debye function showing how the heat capacity due to atomic vibrations varies as a function of the absolute temperature, normalised by the Debye temperature T_D where the heat capacity reaches $3Nk$.

It is convenient to define a new function H, the enthalpy of the system:

$$H = U + PV. \qquad (2.4)$$

A change in enthalpy accounts for both the heat absorbed at constant pressure, and the work done by the $P\Delta V$ term. The specific heat capacity measured at constant pressure is therefore given by:

$$C_P = \left(\frac{\partial H}{\partial T} \right)_P.$$

Heat capacity can be measured using a variety of calorimetric methods. The data can then be used to estimate enthalpy changes as a function of temperature and pressure:

$$\Delta H = \int_{T_1}^{T_2} C_P \, dT. \qquad (2.5)$$

For solids that are not too compressible (i.e. have large bulk moduli), the difference between C_V and C_P is small, simply because the effect of pressure on volume is small. In the case of iron with a face-centred cubic crystal structure, $C_P \approx (1 + 10^{-4}T)C_V$ [7]. For gases, the difference can be large; in the case of air, oxygen, carbon dioxide, and hydrogen, C_P/C_V is found to be 1.4025, 1.3977, 1.2995 and 1.4084, respectively [8].

We have thus far considered only lattice vibrations as contributing to the heat capacity, but electrons make a contribution, albeit relatively small because the Pauli exclusion principles prevents all but those able to change their energies, from participating. Similarly, the paramagnetic to ferromagnetic transition can make a large difference, as illustrated in Figure 2.2 for the case of body-centred cubic iron.

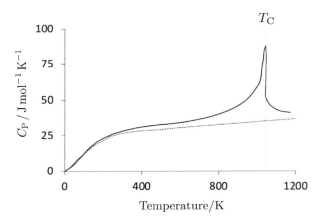

FIGURE 2.2 Specific heat capacity of body-centred cubic iron. The dashed curve represents contributions from lattice vibrations and electrons, whereas the continuous curve representing the total heat capacity, additionally includes magnetic contributions. T_C is the Curie point for the paramagnetic to ferromagnetic transition. Data from [7].

Example 4:

Prove that

$$\left(\frac{\partial U}{\partial T}\right)_P = C_P - P\left(\frac{\partial V}{\partial T}\right)_P$$

Solution 4

From Equation 2.4, $H = U + PV$ so it follows that

$$\left(\frac{\partial H}{\partial T}\right)_P = \left(\frac{\partial U}{\partial T}\right)_P + P\left(\frac{\partial V}{\partial T}\right)_P$$

$$\therefore \left(\frac{\partial U}{\partial T}\right)_P = \underbrace{\left(\frac{\partial H}{\partial T}\right)_P}_{C_P} - P\left(\frac{\partial V}{\partial T}\right)_P$$

2.2.2 Instrument to measure enthalpy

Calorimetric measurements in the distant past involved large instruments known as *bomb calorimeters* [9] to characterise samples weighing 5-25 g. Large calorimeters designed for extreme stability and precision are still used but with small samples, Figure 2.3.

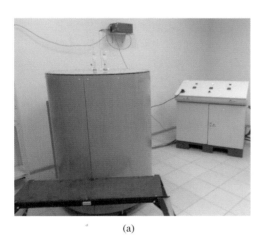

(a)   (b)

FIGURE 2.3   (a) An exceptionally precise, very stable, very large, thermopile calorimeter at Alexandra Khvan's laboratory at NUST-MISIS [10]. A thermopile converts thermal energy into electrical energy. Essentially several thermocouples connected in series. An instrument like this is used in research. (b) Illustration of the sample size used in the thermopile.

However, routine measurements are now possible on very small samples [11]. Differential scanning calorimetry (DSC) is a method for measuring the enthalpy necessary to establish a nearly zero temperature difference between a substance and an inert reference material, as the two specimens are heated or cooled at a controlled rate, Figure 2.4.

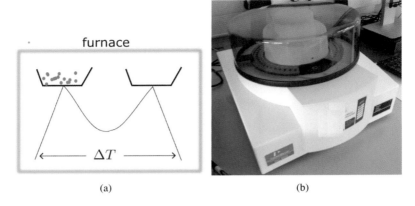

furnace

(a)   (b)

FIGURE 2.4 (a) Schematic of differential scanning calorimeter. (b) Desktop differential scanning calorimeter that can sequentially and automatically characterise many pre-loaded samples.

The temperature difference ΔT depends on whether heat is released or absorbed by the sample relative to the reference. It depends also on whether the ability of the sample to absorb heat from the furnace is different from that of the reference, because of its unalike heat capacity.

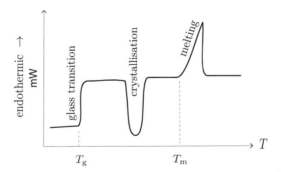

FIGURE 2.5 DSC trace from a sample that initially is glassy, then crystallises and finally, melts during heating. The vertical axis is the heat supplied to the sample, so an exothermic reaction would require the sample to absorb less heat from the furnace, leading to a downward peak.

Figure 2.5 shows the DSC-output during heating at a constant rate, from a sample that initially was in the glassy state with a sudden increase in heat capacity at the glass transition temperature T_g. In the glassy state, the randomly arranged atoms are configurationally frozen so the structure cannot relax. Beyond T_g, the atoms can relax over distances less than an average interatomic spacing, enabling the still randomly arranged atoms to absorb more energy with each increment of temperature, relative to the glassy state. The step in the DSC curve near T_g

reflects this change in heat capacity. There is a significant exothermic peak during crystallisation, associated with the (−ve) "latent heat" of transformation from random to the ordered state that crystals represent. When melting eventually occurs, the latent heat is endothermic.

Transitions from a frozen glassy state to a more relaxed random state are common in polymers but metals can also be made glassy. To achieve the cooling rate necessary to avoid crystallisation in simple alloys, the melt is released on to a spinning, water-cooled copper wheel, with the metallic glass coming off the wheel in the form of a ribbon, as illustrated in Figure 2.6. Guess what such a material could be used for.

(a)   (b)

FIGURE 2.6   (a) Snapshot of molten tin solidifying into a glassy state on the surface of a spinning copper wheel. (b) The metallic glass ribbon that comes off the copper wheel. Images courtesy of Mark Jolly.

Example 5: Metallic glass and differential scanning calorimetry

(i)   Describe how a metal could be forced into a glassy state. How do the properties of a metallic glass vary with the direction in which the sample is tested? Describe, with justification, how such a metal might be used in an electrical application.

(ii)   Figure 2.7 shows a differential scanning calorimetry (DSC) trace during the heating of a sample of an iron-rich metallic glass at $10\,K\,min^{-1}$. Explain the step-like change at about 520 K. What does the inverted peak that begins at about 700 K indicate is happening to the glass? If the sample weight is 0.1 mg, calculate the latent heat per gram, associated with the event beginning at 700 K and ending at 900 K.

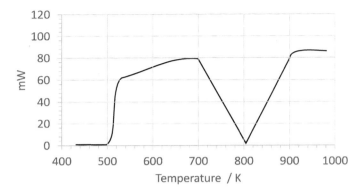

FIGURE 2.7 A DSC curve from a heated sample of metallic glass, with endothermic change increasing along the y-axis.

Solution 5

(i) In order for a material to be a glass, the random atomic-configurations in the liquid must be frozen before crystallisation can occur. In the case of a metal, this often necessitates rapid cooling, for example by allowing liquid metal to pour on to a rotating, water-cooled copper wheel. There are several other methods, the most recent being three-dimensional layer-by-layer printing; a laser is applied to a layer of powder to fuse it, a fresh layer of powder is deposited followed by the laser treatment, with the process repeated until a solid object of the desired shape has evolved. Iron-based metallic glass has been produced successfully in this manner [12]. The rapid heating and cooling associated with the passage of the laser is sufficient to make some alloys glassy.

A glass is isotropic because it has no periodic structure. The average distance between atoms is identical in all directions. Therefore, it would be expected to have the same properties irrespective of the direction of testing.

Since there is no structure, the glass would be magnetically soft, i.e., it would be easy for magnetic domain boundaries to move and reverse direction without much dissipation of energy, Figure 2.8. So, metallic glass is used to make certain transformers where the the field reverses at 50-60 Hz. Such transformers therefore exhibit minimal heating.

domain boundary

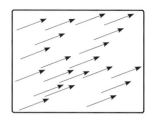

FIGURE 2.8 A single crystal of a ferromagnetic material, in the absence of an external field, spontaneously divides itself into magnetic domains to minimise energy by reducing the extent of its external field. When a field is applied, the domains best aligned to that field grow. The ease of domain boundary motion that facilitates this change defines a *magnetically soft* material. In an electrical transformer, the field changes frequently so a magnetically soft material avoids the hysteresis loss as the domain boundaries move backwards and forwards in unison with the change in field direction. This reduces heat generation as the transformer operates. Glasses do not have structure that can interfere with domain boundary motion.

(ii) The step-like change represents the glass-transition temperature, where atoms can relax so that the volume changes with temperature at a higher rate than when the atoms were configurationally frozen. Therefore, the heat capacity changes but there is no heat evolution or absorption.

The energy emitted (exothermic crystallisation) is given by the area of the peak which is presented as a triangle. Therefore, its area is $\approx 0.5 \times 200 \times 80\,\text{mW K} \equiv \text{mJ s}^{-1}\,\text{K}$. Dividing this by the heating rate which is $10\,\text{K min}^{-1} \equiv 10/60\,\text{K s}^{-1}$ gives 48 J. Dividing this by the mass $(0.1\,\text{mg} \equiv 10^{-4}\,\text{g})$ gives $4.8 \times 10^5\,\text{J g}^{-1}$.

2.3 ENTROPY, FREE ENERGY

In Figure 2.5, melting during heating of the crystallised polymer is associated with an endothermic event, i.e., $H_{\text{melt}} - H_{\text{crystal}} = +\text{ve}$ so the enthalpy change caused an increase in the free energy change ΔG accompanying melting. Since reactions occur spontaneously only if $\Delta G = -\text{ve}$, something else must be driving the melting in a manner that allows the free energy to be reduced in spite of the enthalpy change.

That 'something else' was defined by Clausius during the 19th century as the *entropy* with the symbol S; even if there is no change in enthalpy, a reaction can occur spontaneously and irreversibly in an isolated system if it leads to an increase in entropy, i.e., $\Delta S > 0$. This is because the free energy is a combination of enthalpy and entropy,

$$G = H - TS. \tag{2.6}$$

Entropy is often associated with the degree of disorder – during melting, the highly ordered crystal changes into a liquid with a random arrangement of molecules, leading to an increase in entropy, which when multiplied by $-T$, reduces the free energy.

There are many kinds of disorder, for example, the thermal entropy from the vibration of atoms about their mean positions in a solid, the transition from aligned magnetic spins to randomly pointing spins, and so on. To illustrate entropy, we will consider one type, the configurational entropy. However, we note that from an experimental perspective, a change in entropy can also be measured using calorimetry, via the heat capacity:

$$\Delta S = \int_{T_1}^{T_2} \frac{C_P}{T}\, dT.$$

Example 6: heat capacity of lead

The heat capacity of lead (Pb) is a function of temperature as follows:

$$C_P = 23.6 + 9.8 \times 10^{-3}T \qquad \text{J mol}^{-1}\text{K}^{-1} \qquad \text{for } 300 < T < 600\,\text{K}$$

There are no phase or magnetic transitions in lead over the stated temperature range. Showing the steps in your method, calculate the enthalpy and entropy change when the sample is heated from 300 to 600 K.

Solution 6

If $C_P = \alpha + \beta T$, then the required changes are given by

$$\Delta H = \int_{T_1}^{T_2} C_P\, dT = \int_{T_1}^{T_2} (\alpha + \beta T)\, dT = \left[\alpha T + \frac{1}{2}\beta T^2\right]_{300}^{600} = 8403\,\text{J mol}^{-1}$$

$$\Delta S = \int_{T_1}^{T_2} \frac{C_P}{T}\, dT = \left[\alpha \ln\{T\} + \beta T\right]_{300}^{600} = 19.3\,\text{J mol}^{-1}\,\text{K}^{-1}$$

Example 7: bond pair probability

Explain why ordered crystals containing an equal mixture of 'A' and 'B' atoms become disordered at a sufficiently high temperature. Note: an example of an ordered crystal is equiatomic brass in which the copper atoms are at the corners of the primitive cubic lattice and zinc atoms at the centre of the cube (or vice versa). What is the probability of finding an A atom next to a B atom in a disordered equiatomic solution?

Solution 7

In ordered crystals, A atoms prefer to be next to B atoms: the enthalpy of ordering ΔH is negative. However, there is a decrease in entropy on ordering so $-T\Delta S$ is positive. The latter term dominates at high temperatures, making $\Delta G = G_{\text{ordered}} - G_{\text{disordered}}$ positive and hence favouring disorder, i.e. leading to a random distribution of atoms.

In a random solution, the probability of finding an A atom next to a B atom (and vice versa) in an equiatomic solution is $p_{\text{AB}} = 2x(1 - x) = 0.5$, where x is the concentration as a mole fraction of A and $(1 - x)$ of B. Here x is taken to be the probability of finding an A atom and $(1 - x)$ of a B atom. The factor of two comes in because we must count both A-B and B-A bonds.

There is no case where an isolated, ordered crystal becomes disordered on *cooling*. The ordered crystal would be in a state where $\Delta H = H_{\text{ordered}} - H_{\text{disordered}}$ is negative so this would oppose disordering. Although entropy favours disorder, its contribution to free energy diminishes linearly with temperature. So if a crystal already is ordered, then there will be no tendency for it to disorder on cooling.

Some ordered protein crystals do become less ordered on cooling, but this is because they contain water which on freezing damages the crystals [13]. Observations dating back to the 1890s noted that some naturally-occurring crystalline minerals became amorphous over time, but this is caused by radiation damage [14].

2.4 CONFIGURATIONAL ENTROPY

Figure 2.9a shows a mixture of two kinds of atoms, with like atoms segregated with no mixing; there is only one way of achieving this arrangement. On the other hand, if they are allowed to mix ideally then there are many more ways of configuring them, three of which are illustrated in Figure 2.9c-d. A mixing of the atoms is obviously more *probable*.

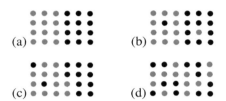

FIGURE 2.9 Four different configurations of a mixture of two kinds of atoms. (a) The two kinds of atoms are partitioned into their own spaces, without mixing. (b-d) If the atoms are allowed to mix then many more arrangements are possible, here only three of the many are illustrated.

Suppose there are N sites amongst which are distributed n atoms of type A and $N - n$ of type B, Figure 2.10. The first A atom can be placed in N different ways and the second in $N - 1$ different ways. These two atoms cannot be distinguished so the number of different ways of placing the first two A atoms is $N(N - 1)/2$; similarly, for placing the first three A-atoms, the distinguishable configurations is $N(N - 1)(N - 2)/3!$. The number of distinguishable ways of placing all the A atoms is

$$\frac{N(N - 1) \dots (N - n + 2)(N - n + 1)}{n!}$$

Since

$$N(N - 1) \dots (N - n + 2)(N - n + 1) \underbrace{(N - n)(N - n - 1) \dots (1)}_{(N-n)!} = N!$$

It follows that

$$\frac{N(N - 1) \dots (N - n + 2)(N - n + 1)}{n!} = \frac{N!}{n!(N - n)!}. \tag{2.7}$$

So if the atoms behave ideally, i.e., they do not have a preference for the type of neighbour, then the probability of a uniform distribution is much much more likely than the ordered distribution.

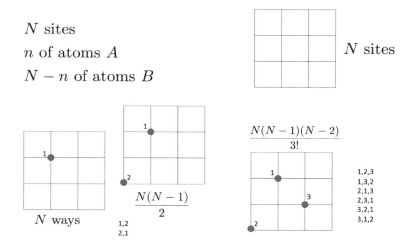

N sites

n of atoms A

$N - n$ of atoms B

N sites

N ways

$\dfrac{N(N-1)}{2}$

1,2
2,1

$\dfrac{N(N-1)(N-2)}{3!}$

1,2,3
1,3,2
2,1,3
2,3,1
3,2,1
3,1,2

FIGURE 2.10 Schematic illustration of the placing of n A-atoms on a lattice with N sites. The first can be placed on any site, but the second only on $N - 1$ sites. However, atoms 1 and 2 cannot be distinguished because an arrangement 1,2 is identical to 2,1. When the third atom is placed, there are 3! indistinguishable ways. The process is complete when $N - n + 1$ A-atoms have been located, with the remainder being filled by B-atoms.

For a real system for which the number of atoms is very large, a parameter is needed that expresses the likelihood as a function of the correspondingly large number of configurations (w_c) possible. Suppose that a term S is defined such that $S \propto \ln w_c$, where the logarithm is taken because it may be necessary to add two different kinds of disorder (after Boltzmann [15, 16]), then the S is identified as the *configurational entropy* $S = k \ln w_c$, where k, the proportionality constant, is known as the Boltzmann constant which for a mole of atoms is the gas constant R.

A simple example illustrates this. Suppose there are two identical particles, each with $N = 10^5$ atoms of which $n = 100$ are A-atoms and the rest B-atoms. The number of arrangements possible in each particle is therefore

$$w_c = \frac{10^5!}{10^2!(10^5 - 10^2)!} = 1.02 \times 10^{342} \tag{2.8}$$

and $\ln\{1.02 \times 10^{342}\} = 788.$

If the two particles are now combined,

$$w_{c,\text{total}} = \frac{2 \times 10^5!}{2 \times 10^2!(2 \times 10^5 - 2 \times 10^2)!} = 1.85 \times 10^{685} \tag{2.9}$$

and $\ln\{1.85 \times 10^{685}\} = 1578$

so it is evident that if we simply add the number of configurations possible for each particle, i.e. $1.02 \times 10^{342} + 1.02 \times 10^{342} \neq 1.85 \times 10^{685}$ where this last number is the configurations possible if we combine the two particles into one. However, if the logarithms of w_c are added, $788 + 788 = 1576$ which is only slightly different from 1578, but this difference diminishes as N becomes large, bearing in mind that a mole contains some 6.023×10^{23} atoms. The entropy is a thermodynamic function of state and so it is additive, Figure 2.11.

$$S = k \ln\{w_c\}$$

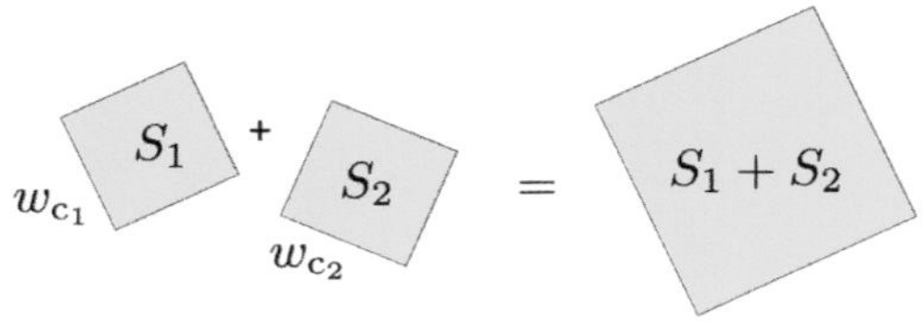

$$w_{c_1} + w_{c_2} \neq w_{c,\text{total}}$$
$$\ln\{w_{c_1}\} + \ln\{w_{c_2}\} = \ln\{w_{c,\text{total}}\}$$

FIGURE 2.11   Entropy is a function of state so the individual entropies of separate particles can be summed to give the total entropy of the combined particle.

When comparing scenarios, the one that is favoured on the basis of the degree of disorder is that which has the greater entropy. In terms of solutions, entropy favours mixing over separation. On this basis, it can be shown quite simply that the change in entropy when atoms mix is given by

$$\Delta S = -R \sum_{i=1}^{j} x_i \ln\{x_i\} \tag{2.10}$$

where $i = 1 \ldots j$ represents the atomic species and x_i its mole fraction. Notice that this configurational entropy of mixing *always* is positive.

Example 8: entropy of mixing

Prove the form of Equation 2.10.

Solution 8

Consider three identical objects, 1,2,3. There are $3! = 6$ ways of arranging them: 123, 132, 213, 231, 321, 312. However, if two of the objects are the same, then the number of ways decreases to $3!/(1!2!) = 3$. Say the 2 and 3 are identical, then the possibilities are 122, 212, 221. We will use this to derive the relationship that follows.

Consider a random mixture of a mole of atoms with a concentration $(1 - x)$ of A atoms and x of B atoms so the total number of atoms is given by Avogadro's number N_a. The number of possible arrangements is

$$w_c = \frac{N_a!}{(N_a(1-x))! \, (N_a x)!} \qquad (2.11)$$

For large numbers, $\ln\{y!\} = y \ln\{y\} - y$ so taking logarithms of Equation 2.11 gives

$$\ln\{w_c\} = [N_a \ln\{N_a\} - N_a] - [N_a(1-x) \ln\{N_a(1-x)\} - N_a(1-x)]$$
$$- [N_a x \ln\{N_a x\} - N_a x]$$

$$= N_a \ln\{N_a\} - N_a - N_a(1-x)\ln\{N_a\} - N_a(1-x)\ln\{1-x\}$$
$$+ N_a(1-x) - N_a x \ln\{N_a\} - N_a x \ln\{x\} + N_a x$$

$$= -N_a[(1-x)\ln\{1-x\} + x\ln\{x\}] \qquad (2.12)$$

Notice that the derivation of equation 2.12 relies on a random mixture of the two kinds of atoms. Such a mixture can only arise in three circumstances:

(i) when the temperature of the mixture is so high that thermal agitation randomises the mixture;

(ii) when individual atoms are indifferent to their neighbours, i.e., when there is no change in energy when breaking an A-A bond and a B-B bond to make two A-B bonds;

(iii) when the material is produced by severely deforming a mixture of the elemental powders of A and B at a temperature where they are unable to diffuse. The deformation eventually breaks the particles into ever decreasing size until the atoms of A and B are mixed together at random [17]. This is the process of mechanical alloying [18].

Example 9: shape of free energy curve

Equation 2.10 gives the entropy of mixing so the free energy of mixing is $\Delta G_M = -T\Delta S_M$. Differentiate this ΔG_M to show that on a plot of free energy versus mole fraction, the slope of the curve should be $\pm\infty$ at $x = 0$ or $x = 1$. In practice, the slope will be large but finite – why is that?

Solution 9

The differential of ΔS_M with respect to x is proportional simply to $\ln\{1-x\}-\ln\{x\}$ which gives $\pm\infty$ at $x = 0$ or $x = 1$. However, this assumes that concentration is continuous whereas it is in fact discrete since concentration is changed by adding or removing an atom or atoms. Therefore, although the slope will be very large, it will not be infinite [19]. One consequence is the care required when drawing free energy curves, Figure 2.12. Another is that the shape of the curve explains why it is so difficult to achieve extreme purity, because the free energy increases sharply as the solute concentration reaches very small concentrations.

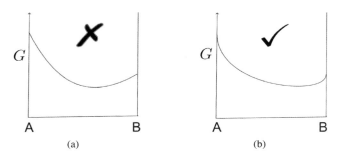

(a) (b)

FIGURE 2.12 The diagram on the left is wrong because the free energy curve should intersect the vertical axes of the pure elements at a slope $\pm\infty$.

Example 10: solutions, solubility, entropy

1. Define what is meant by an ideal solution. For a binary A-B ideal-solution, what would be the slope of the curve representing the free energy of mixing versus concentration, at the pure A and pure B axes? Based on your answer, comment on why it is very difficult to purify materials.
2. Sometimes, solutions have very limited solute-solubility. How could you force such atoms to mix and form a random solution with a solute concentration that is well above the solubility limit?
3. Atoms of A and B are arranged in a straight line at random, with the mole fraction of B equal to x. What is the probability of finding two A atoms next to each other? How would your calculation be modified if this was a two-dimensional array of A and B atoms?
4. An alloy is to be made, containing Fe, Mn, Al, Au and Mg. What are the concentrations of these elements that result in a maximum configurational entropy of mixing?

Solution 10

(i)   The enthalpy of mixing is zero for an ideal solution, i.e., there is no change in bond energy when A-A and B-B bonds are broken to form 2A-B bonds. The atoms in such a solution will be randomly located at all temperatures. The configurational entropy of mixing ΔS_M, and hence the free energy of mixing ΔG_M has a term of the form $x \ln\{x\} + (1-x) \ln\{1-x\}$, which when differentiated with respect to x gives $\ln\{1-x\} - \ln\{x\}$ – this in turn gives $\pm\infty$ at $x = 0$ or $x = 1$. This means that the free energy changes sharply at concentrations close to the pure elements when small concentrations of impurities are added, making purification difficult.

(ii)   To force elements to mix even when they do not prefer to mix, mechanical forces are used. The elemental powder or elemental compounds are put into a ball mill containing hard balls that severely deform the powders, forcing mixtures to form on an atomic scale.

(iii)   In a straight line, with a random arrangement, the probability is simply the concentration of the element concerned, squared since the wish is to discover near-neighbour pairs of like-atoms. In a two-dimensional array, this would need to be multiplied by the coordination number z because pairs can be formed along different directions.

(iv)   The configurational entropy for any number of solutes involved, maximises when their atomic concentrations are set to be equal.

2.5 HIGH-ENTROPY ALLOYS

Equation 2.10 indicates that in ideal solutions, an equiatomic mixture of five elements would have an entropy of mixing of $\Delta S = 1.61R$ whereas for an equiatomic binary solution, $\Delta S = 0.69R$. This is the basis of the so-called high-entropy alloys [20, 21] where by maximising the entropy of mixing, the tendency to precipitate phases is reduced and likewise, the tendency for a multicomponent, concentrated mixture to form a single solid-solution is increased. One example is the CrMnFeCoNi equiatomic alloy which solidifies as a single face-centred cubic solid solution, even though it is only nickel which has the face-centred cubic structure under ambient conditions.

The concept can be extended to ceramics [22]. In a thermoelectric material, a temperature gradient leads to an electrical potential, with the ratio of the latter to the former designated the *Seebeck coefficient*. A large value of this coefficient increases the utility of the thermoelectric material, but a low thermal conductivity is also required so that the temperature gradient can be maintained. This is the basis of the electricity source on satellites, where a plutonium heat source is surrounded by a SiGe thermoelectric generator (Figure 2.13).

A novel high-entropy perovskite ceramic $Sr_{0.9}La_{0.1}(Zr_{0.25}Sn_{0.25}Ti_{0.25}Hf_{0.25})O_3$ prepared using ball milling of powders exhibits both a low thermal-conductivity and a large Seebeck coefficient, Figure 2.13b [22]. The conductivity is reduced because several types of atoms occupy the same lattice site at random, which disrupt the thermal vibrations responsible for heat flow.

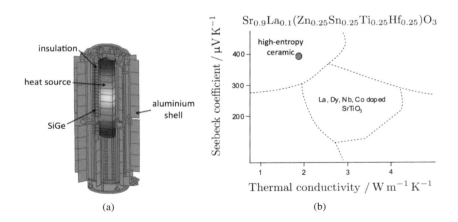

(a)   (b)

FIGURE 2.13 (a) A thermoelectric battery for satellite missions to other planets, where the heat is generated using a radioactive source and the resulting temperature gradient induces a potential difference across the SiGe to provide electricity. Image courtesy of Parker Betty, adapted under the CC-BY-SA-4.0 licence https://creativecommons.org/licenses/by-sa/4.0/deed.en (b) The red dot identifies the low thermal conductivity, high Seebeck coefficient high-entropy ceramic. Plot simplified from [22].

Chapter 3

Equilibrium state

In a single-phase equilibrium diagram such as that for iron as a function of temperature and pressure, the boundaries between the phase fields represent the locus of all points along which the adjacent phases are in equilibrium, i.e., they have an identical free energy. For example, the α/γ equilibrium is defined by setting (Figure 3.1):

$$G^{\alpha} = G^{\gamma}. \tag{3.1}$$

This is because allotropic transitions are considered here as a function of variables such as temperature and pressure, where the crystal structure changes but not the chemical composition – an allotrope is just a different physical form of the same element.

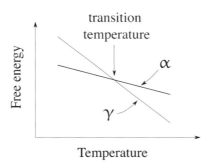

FIGURE 3.1 The transition temperature for an allotropic transformation. In the case of a pure substance, equilibrium is when the phases have identical free energy.

The phase diagram for pure iron as a function of both temperature and pressure is shown in Figure 3.2, illustrating the solid phases α, γ, ϵ and δ, together with its liquid state. Each of the lines indicates an equilibrium between two phases, for example, the δ/L line is the locus of all points where $G_{\delta} = G_{L}$. There clearly exists a combination of temperature and pressure where all three of α, γ, ϵ can coexist in equilibrium, with $G^{\alpha} = G^{\gamma} = G^{\epsilon}$. At very high pressures and quite high temperatures, the densest phase ϵ is most stable – it is believed therefore that this is the solid phase weighing some 10^{23} kg that exists at the core of the Earth. Experiments that expose the core to longitudinal and transverse shock waves seem to support this conclusion.

Calculations indicate that at much greater pressures than plotted on Figure 3.2, and pressures expected at the cores of exoplanets, other allotropes of iron may exist, including a body-centred tetragonal crystal structure that is not ferromagnetic [23].

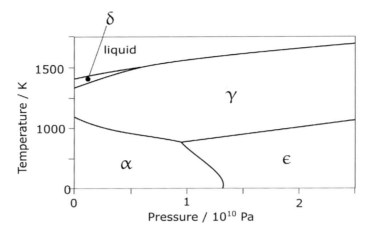

FIGURE 3.2 Temperature versus pressure equilibrium phase diagram for pure iron. The triple point temperature and pressure are 490°C and 11 GPa respectively. α, γ and ε refer to ferrite, austenite and ε-iron respectively. Diagram courtesy of Shaumik Lenka.

The phase diagram for pure carbon is shown in Figure 3.3 based on both experimental and calculated data. The manufacture and use of synthetic diamonds has increased sharply, both in the industrial and jewellery markets – the phase diagram defines the conditions under which diamond can be produced. Under all conditions, diamond has a greater density than graphite or liquid carbon, so it is not surprising that it becomes the most stable phase a high pressures. It is, however, the case that at pressures greater than 500 GPa, liquid carbon becomes more dense than cubic diamond, in which case the liquid/diamond phase boundary on a temperature versus pressure plot assumes a negative slope [24].

There are many other allotropic forms of carbon, for example, diamond with a hexagonal crystal structure [25], but these are not incorporated on equilibrium phase diagrams, presumably because cubic-diamond and graphite are the most stable forms. There are claims that carbon nanotubes can be transformed into minute diamonds at appropriate pressures and temperatures.

The planets Uranus and Neptune have about a fifth of their mass in the form of carbon and pressure-temperature conditions are such that compounds such as methane would dissociate into pure carbon and hydrogen. This carbon is assessed to be converted into diamond (or alloy with metallic hydrogen), so these planets have been referred to as being "diamonds in the sky" [26].

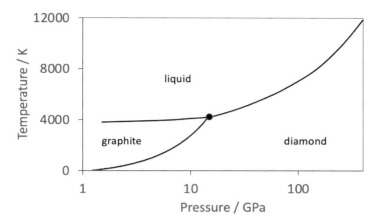

FIGURE 3.3 Temperature versus pressure equilibrium phase diagram for pure carbon, excluding the vapour phase. The diamond here has a cubic crystal structure. The maximum pressure plotted is 395 GPa. Data from Ghiringhelli et al. [27].

3.1 CHEMICAL POTENTIAL

A different approach is needed when the chemical composition is a variable. Consider a single-phase alloy consisting of two components A and B. The molar free energy $G\{x\}$ of that phase will in general be a function of the mole fractions $(1-x)$ and x of A and B respectively, written as a weighted mean of the free energy contributions from each component:

$$G\{x\} = \underbrace{(1-x)\mu_A}_{\text{contribution from A atoms}} + \underbrace{x\mu_B}_{\text{contribution from B atoms}} . \qquad (3.2)$$

The terms μ_A and μ_B, known as the *chemical potentials* per mole of A and B respectively, in effect partition the total free energy $G\{x\}$ into a component purely due to A atoms and another due to B atoms alone. This equation is illustrated in Figure 3.4 by the tangent at the coordinate $[G\{x\}, x]$. Consistent with Equation 3.2, the intercepts of this tangent on the vertical axes give μ_A and μ_B. Since the slope of the tangent depends on the composition, so do the chemical potentials. Note that the free energies of the pure components are written μ_A° and μ_B°.

It is evident from Figure 3.4 that

$$\mu_A = G\{x\} - x\frac{\partial G}{\partial x}$$

and

$$\mu_B = G\{x\} + (1-x)\frac{\partial G}{\partial x}$$

where $\partial G/\partial x$ is the slope of the tangent so the product on the right-hand side of the equations simply represents the difference in μ and G.

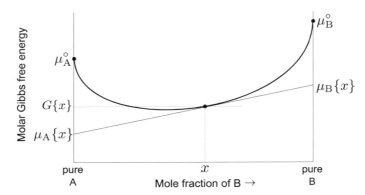

FIGURE 3.4 Illustration of the chemical potential μ for a binary solution, with μ° representing the free energy of the pure component. $G\{x\}$ is the molar Gibbs free energy of a mixture of composition x, which is the mole fraction of B.

The chemical potential $\mu\{x\}$ of a component is known also as its *partial* molar free energy, describing a part of the *integral* molar free energy $G\{x\}$. There are in fact many quantities which can be expressed using relationships of the form implied by Equation 3.2. Thus, the volume of a solution might be written in terms of the partial molar volumes of the components:

$$V_{\mathrm{m}} = \overline{V}_{\mathrm{A}} x_{\mathrm{A}} + \overline{V}_{\mathrm{B}} x_{\mathrm{B}} \tag{3.3}$$

where $\overline{V}_i$ refers to the partial molar volume of component i =A,B.

3.2 EQUILIBRIUM BETWEEN SOLUTIONS

Consider now two phases α and γ that are placed in intimate contact in a binary steel. The phases will only be in equilibrium with each other if the carbon atoms in γ have the same chemical potential as the carbon atoms in α, and if this is true also for the Fe atoms:

$$\begin{aligned} \mu_{\mathrm{C}}^{\alpha} &= \mu_{\mathrm{C}}^{\gamma} \\ \mu_{\mathrm{Fe}}^{\alpha} &= \mu_{\mathrm{Fe}}^{\gamma}. \end{aligned} \tag{3.4}$$

In fact, in a binary solution, the chemical potentials of A and B when sharing a tangent are not independent so this last condition is redundant. This is apparent from Figure 3.4, where the two potentials are connected by the tangent.

If the atoms of a particular species have the same chemical potential in both the phases, then there can be no tendency for them to migrate across the phase boundaries. The system will be in stable equilibrium if this condition applies to all species of atoms. The way in which the free energy of a phase varies with concentration is unique to that phase, so the *concentration* of a particular species of atom need not be identical in phases which are at equilibrium. Therefore, in general,

$$x_C^{\alpha\gamma} \neq x_C^{\gamma\alpha}$$
$$x_{Fe}^{\alpha\gamma} \neq x_{Fe}^{\gamma\alpha} \tag{3.5}$$

where $x_i^{\alpha\gamma}$ describes the mole fraction of element i in phase α which is in equilibrium with phase γ etc.

The condition that the chemical potential of each species of atom must be the same in all phases at equilibrium is general. For the binary alloy, two phase case, it follows that the equilibrium compositions can be found on a plot of free energy versus composition, by constructing a tangent that is common to the two free energy curves as illustrated in Figure 3.5.

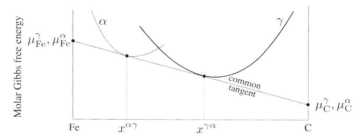

FIGURE 3.5   The common tangent construction giving the equilibrium compositions $x^{\alpha\gamma}$ and $x^{\gamma\alpha}$ of the two phases at a fixed temperature.

Example 11: definition of equilibrium

Giving reasons, explain which of the following statements are incorrect:

(a) Equilibrium between two phases in a pure substance is given by the temperature at which they have equal free energies.
(b) Equilibrium between two phases in a binary alloy occurs when the phases have an equal free energy.
(c) It is not possible for three phases to be in equilibrium at the same time.
(d) The chemical compositions of two phases in a binary system at equilibrium will not be identical.

Solution 11

(b) is not correct. For a binary or multicomponent system, the chemical potentials of each element must be uniform across the phases. The free energies of the phases can be unequal. It the chemical potentials are equal, there is no tendency for the migration of a species across the phases.

(c) is not correct – any number of phases can coexist in equilibrium, the general condition being that the chemical potential of each species is uniform across the phases.

3.3 PHASE DIAGRAMS AND FREE ENERGY SURFACES

The way in which free energies and the common tangent construction can be used to construct phase diagrams is illustrated on Figure 3.6, for the simple case where there is complete solid-solubility and liquid-solubility over the entire range of compositions of molybdenum and titanium. At temperatures above the melting point of Mo, the solid free energy is above that of the liquid for all compositions, so only liquid becomes the stable phase. Similarly, below the melting temperature of titanium the alloy is solid for all compositions.

At an intermediate temperature, there is a region where both solid and liquid, albeit of different chemical compositions, are stable. The free energy curves of the solid and liquid intersect so it is possible to draw a common tangent that defines the compositions of the solid and liquid phases that are in equilibrium between the points marked. Between the points, a mixture of solid and liquid has a lower free energy than either of the single phases.

If the mass fraction is represented by m_f then within the two-phase field, it follows from mass balance that the average composition of the alloy ($\bar{x}$) must equate the weighted average of the solid and liquid, Figure 3.7:

$$\bar{x} = m_f^S x_S + m_f^L x_L$$
$$\equiv m_f^S x_S + (1 - m_f^S)x_L$$

so that $\quad m_f^S = \dfrac{\bar{x} - x_L}{x_S - x_L} \quad$ and $\quad m_f^L = 1 - m_f^S \qquad$ (3.6)

On Figure 3.7, consistent with Equation 3.6,

$$m_f^S \approx \frac{bc}{ac} \quad \text{and} \quad m_f^L \approx \frac{ab}{ac}, \qquad (3.7)$$

the approximation sign is used here because the units of concentration are not expressed in terms of mass; the 'bc' etc. refer to the distances between the labelled points on Figure 3.7; the line 'abc' that connects the equilibrium compositions of the phases is known as a tie-line. Obtaining phase fractions as in Equation 3.6 is known as the *lever rule*.

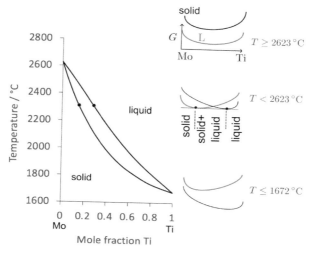

FIGURE 3.6 A binary phase diagram with illustrative free-energy curves on the right. At intermediate temperatures between the melting temperatures of pure Ti and Mo, it is possible to define a two-phase field where solid and liquid of different chemical compositions are in equilibrium. Both molybdenum and titanium under these conditions have a body-centred cubic crystal structure. There is complete mutual solubility of these elements in each other. NiO and MgO, both of which have the same crystal structure (cubic-F with a motif of a metal atom at 0,0,0 and oxygen atom at $\frac{1}{2}$, 0, 0 associated with each lattice point), exhibit the same behaviour, with solid-solubility and liquid-solubility over the entire range of compositions. Copper and nickel, both of which have the cubic close-packed crystal structure, also form a phase diagram such as this. If you are interested in crystal structure, download free book from: https://www.phase-trans.msm.cam.ac.uk/2020/Crystallography_book.pdf

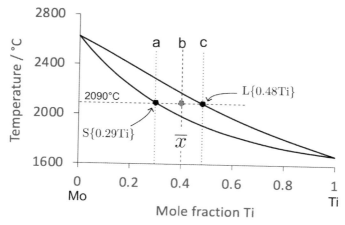

FIGURE 3.7 The red dot represents the average chemical composition $\overline{x} = 0.4$ of an alloy that has been cooled to 2090 °C within the liquid+solid phase field. S and L represent solid and liquid, respectively. Liquid of composition $x_L = 0.48$ then exists in equilibrium with solid of composition $x_S = 0.29$.

3.4 EUTECTIC

Phase diagrams often contain more than one kind of solid phase; for example, in the lead-tin system there are the α and β solid phases at low temperatures. These are lead-rich and tin-rich solid solutions, respectively. α has a face-centred cubic crystal structure whereas β-tin has a tetragonal lattice.

Lead-tin alloys, because they melt at relatively low temperatures, are often used as connectors or solders in a variety of electronic and battery applications. Ancient *Pewter*, a tin-rich alloy containing about 30 wt% of lead, was used for food containers, Figure 3.8, but because lead can be toxic, modern versions contain alternative alloying additions to tin.

FIGURE 3.8 Pewter utensils at the Edo Museum in Tokyo. These may date to the Shogun era, about 130 years ago.

A low melting-temperature is obviously beneficial when soldering, and it also is important that the solder solidifies at a well-defined temperature because otherwise, the joint may be disturbed while solidification progresses over a range of temperatures. Lead-tin alloys have a particular advantage in both these respects, because the melting temperature of the alloy can be much less than that of the pure components – the phase diagram exhibits a *eutectic* at 183 °C and 62 wt% Sn, Figure 3.9. This means that a liquid of that composition solidifies at 183 °C, to generate a mixture of two solid solutions, i.e., $L \rightarrow \alpha + \beta$, identified as the eutectic reaction. If the tin concentration is less than $\bar{x}_{Sn} < 62$ wt% then the alloy is said to be *hypoeutectic* with solidification beginning by the precipitation of α, where $\bar{x}_{Sn}$ is the average composition of the alloy. Since $x_{Sn}^{\alpha L} < \bar{x}_{Sn}$, the excess tin is partitioned into the liquid, causing its composition to enrich in Si. With the continued precipitation of α as the temperature decreases, the liquid composition eventually reaches the eutectic composition and it decomposes into the eutectic mixture of $\alpha + \beta$ at 183 °C.

Alloys in which $\bar{x}_{Sn} > 62$ wt% are designated *hypereutectic* and begin solidification with the precipitation of β until the liquid that is enriched in this process with Pb, decomposes into the eutectic mixture of α+β at 183 °C. In summary,

$$\bar{x}_{Sn} = 62 \text{ wt\%}, \ T = 183°C \qquad L \rightarrow \underbrace{\alpha + \beta}_{\text{eutectic}}$$

$$\bar{x}_{Sn} < 62 \text{ wt\%}, \ T > 183°C \qquad L \rightarrow L' + \underbrace{\alpha}_{\text{proeutectic}}$$

$$\downarrow$$

$$T = 183°C \qquad \underbrace{\alpha + \beta}_{\text{eutectic}}$$

where L' is the liquid that has changed in composition due to the formation of proeutectic α. It is seen that both proeutectic and eutectic α will be present in the microstructure of a hypoeutectic alloy. A similar rationale applies to hypereutectic alloys.

$$\bar{x}_{Sn} > 62 \text{ wt\%}, \ T > 183°C \qquad L \rightarrow L' + \underbrace{\beta}_{\text{proeutectic}}$$

$$\downarrow$$

$$T = 183°C \qquad \underbrace{\alpha + \beta}_{\text{eutectic}}$$

The interpretation of the phase diagram in terms of free energy curves is shown in Figure 3.9 for temperatures above, at and below the eutectic temperature. The common tangents identify the compositions of two or three phases that are in equilibrium at the specified temperature – it is at the eutectic temperature that liquid, α and β are in three-phase equilibrium.

It also is possible to have a *eutectoid* reaction in which a solid phase $\gamma \rightarrow \alpha + \beta$ so there is no liquid involved. Such a reaction will be dealt with in the context of the Fe-C phase diagram.

In the case of the Pb-Sn phase diagram, there is some solubility of tin in α and of lead in β. However, with aluminium-silicon alloys, the silicon-rich solid solution has negligible solubility for aluminium when in equilibrium with liquid or with the aluminium solid solution. As a result the eutectic phase diagram in Figure 3.10 looks somewhat different at the silicon-rich side because the silicon solid solution contains a very small concentration of aluminium, making that phase field so narrow that it appears absent. Al-Si alloys are particularly important in the manufacture of blocks for internal combustion engines because of the low melting temperature, high thermal conductivity and a good strength-to-weight ratio.

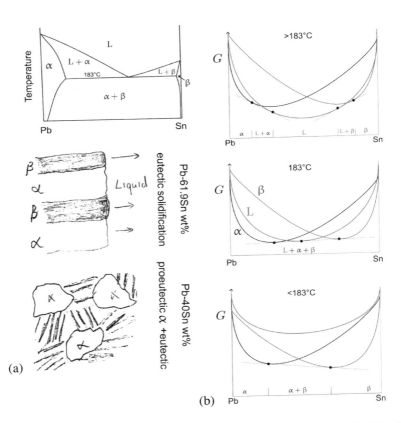

FIGURE 3.9 (a) The lead-tin equilibrium phase diagram. α and β are lead-rich and tin-rich solid solutions, respectively. The eutectic composition has 61.9 wt% of tin and solidification occurs by the simultaneous formation of alternative layers of both α and β as they grow with a common front with the liquid. When the composition contains less tin, the first phase to solidify is proeutectic α which enriches the remaining liquid in tin until it is able to decompose by a eutectic reaction. (b) Disposition of the free energy curves as a function of the temperature about the eutectic temperature. The common tangents identify the phases present in equilibrium.

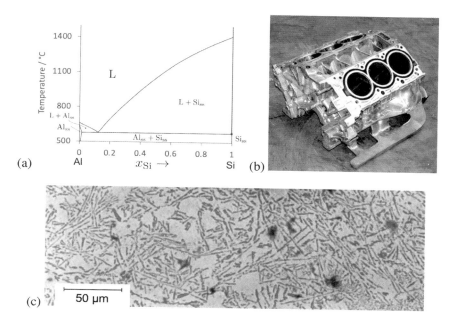

FIGURE 3.10 (a) The aluminium-silicon equilibrium phase diagram. The subscript 'ss' in Al_{ss} and Si_{ss} indicates a solid solution. (b) Cast V6 engine block made from Al-7Si-3Mg-0.5Mn-0.25Zr wt%. Image courtesy of the Institute of Cast Metals Engineers. (c) Eutectic microstructure of an Al-12Si wt% alloy.

Example 12: eutectic mixtures

(i)   Why does a eutectic mixture grow rapidly from the liquid, when compared with a proeutectic phase precipitating from the liquid phase?

(ii)   Giving reasons, explain how you could attempt to produce a eutectic mixture that has a fine interlamellar spacing?

(iii)   Why are alloys that transform into a eutectic mixture used in the manufacture of (a) blocks for internal combustion engines; (b) utensils made of cast metal in ancient times? In each case, state the elements that are used to make the alloy. Give an example where a eutectoid mixture is used in transportation.

Solution 12

(i)   A eutectic has the same average chemical composition as the parent liquid, so the liquid phase does not become enriched as the transformation proceeds. This is not the case with a proeutectic phase, the composition of which is different from that of the liquid, which therefore becomes enriched or depleted with solute, causing the reaction to slow as it progresses.

(ii) Since interfaces between the phases contained within the eutectic are created to a larger extent when the structure is fine, the transformation must be induced by supercooling the liquid to a lower temperature. This means that the driving force for the transformation is enhanced, so that a greater surface area per unit volume of interfaces can be accommodated.

(iii) Eutectics are useful as casting alloys because the melting temperature is at a minimum. Engine blocks are Al-Si, and the ancient eutectic is pewter (Pb-Sn). Rails are made from a eutectoid mixture of cementite and ferrite that grow from austenite.

Example 13: solidification of Al-Cu

An alloy of composition 90Al-5Cu wt% is slowly solidified. The microstructure obtained is illustrated in Figure 4.2, along with a phase diagram of the Al-Cu system.

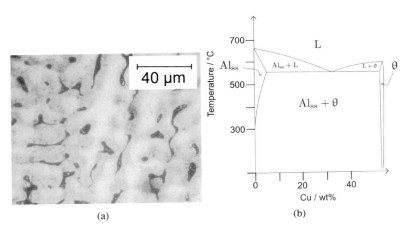

(a)   (b)

FIGURE 3.11 (a) Microstructure of solidified alloy. Image courtesy of T. W. Clyne under CC BY-NC-SA 2.0 UK license, DoITPoMS micrograph library [28]. (b) Part of the Al-Cu phase diagram.

(i) Sketch the important features of the microstructure illustrated, indicating the phases present.

(ii) Explain the development of the microstructure during cooling from the liquid to room temperature.

(iii) Estimate the fraction of the alloy expected to have the eutectic composition. How does this compare with the micrograph shown?

(iv) Give a rough estimate of the temperature needed to homogenise the microstructure in 800 s, given that the interdiffusion coefficient is $D = 5 \times 10^{-6} \exp\{-120,000/RT\} \, \text{m}^2 \, \text{s}^{-1}$ with the gas constant $R = 8.3143 \, \text{J} \, \text{mol}^{-1} \, \text{K}^{-1}$.

Solution 13

(i) The light areas represent the primary Al-rich phase labelled Al_{ss} on the phase diagram, with the darker areas representing the eutectic mixture of Al_{ss} and θ.

(ii) The primary Al_{ss} is expected to solidify at around 630 °C. The first solid to form is much less Cu-rich than the liquid ($\approx 1.5 \, \text{wt}\%$). As solidification proceeds, the liquid becomes progressively richer in Cu, but because diffusion may not occur at a rate consistent with equilibrium, so the solid is cored. When the eutectic temperature is reached the remaining liquid forms the eutectic.

The primary Al_{ss} has developed as dendrites because of constitutional supercooling.

(iii) The eutectic composition is about 32 wt% Cu, so using the lever rule, the mass fraction of eutectic expected under equilibrium conditions would be $(10 - 6)/(32 - 6) = 0.15$. This will not compare exactly with the micrograph, which may have a larger quantity of eutectic because of the fact that the Al_{ss} is cored.

(iv) The coring is over a distance roughly 10 μm, so taking this as the diffusion distance z, and assuming random walk, $z = \sqrt{6Dt}$, $D = z^2/(6t) = 2.08 \times 10^{-14} \, \text{m}^2 \, \text{s}^{-1}$. Setting this equal to the diffusion coefficient, yields the required temperature as 748 K.

3.5   TERNARY AND HIGHER ORDER PHASE DIAGRAMS

In a ternary alloy, there are three concentration variables and temperature (or pressure ...). We have to draw a three-dimensional diagram with temperature on the vertical axis and of a triangular cross section to represent the three components. An example of an isothermal section of a part of the Ni-Al-Ti system is illustrated in Figure 3.12. The lines within the two-phase field are tie-lines indicating compositions of phases at either end that are in thermodynamic equilibrium. Similarly, the corners of the three-phase fields represent equilibrium compositions.

Three-dimensional phase diagrams are obviously difficult to visualise either

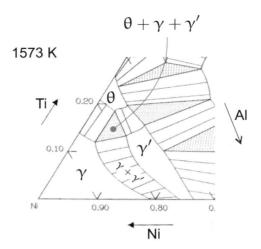

$\theta + \gamma + \gamma'$

1573 K

Ti

0.20

θ

Al

0.10

γ'

γ

Ni

0.90

0.80

0.

Ni

FIGURE 3.12 An isothermal cross-section of the nickel-titanium-aluminium system.

on paper or on a computer screen; information is limited in two-dimensional vertical or isothermal sections. And most practical alloys contain many more elements, typically 10-20. It is then impossible to visualise so many dimensions. The relevant question then becomes: what information do we get from any phase diagram, given a chemical composition of a material and the temperature?

(i) the phases that are in thermodynamic equilibrium;
(ii) the chemical compositions of each of the phases;
(iii) the proportion of each phase.

Given the large quantity of thermodynamic data available, it is possible, using commercially available software and databases, to calculate these quantities for multicomponent systems. Figure 3.13 below illustrates a calculation for a seven element system which has three phases in equilibrium for the temperature and pressure studied. Calculations like these are now routine in materials science, both in industry and academia. It therefore is not necessary to draw multicomponent phase diagrams.

```
Temperature =    743.0000 K

Fixed pressure =  1.013250E+05 Pa,    1.000000E+00 atm

Component     Ref.Phase      Chem.Pot.              Amount/mol   Mass/kg
Fe                          -2.680008E+04           1.732143E+03 9.673500E+01
C                           -4.715406E+02           8.658730E+01 1.040000E+00
Si                          -1.649198E+05           8.901390E+00 2.500000E-01
Mn                          -6.900313E+04           6.370818E+00 3.500000E-01
Ni                          -6.255415E+04           2.129835E+00 1.250000E-01
Mo                          -8.212809E+04           5.211591E-01 5.000000E-02
Cr                          -5.468734E+04           2.788676E+01 1.450000E+00
Total                                               1.864540E+03 1.000000E+02

Amount      Phase              Mole fraction of component within phase
compnt moles
                               Fe            C             Si
1.5174E+03 BCC_A2            0.9918499     0.0000230     0.0058664
3.4154E+02 CEMENTITE         0.6555259     0.2500000     0.0000000
5.6465E+00 M23C6             0.5784515     0.2068966     0.0000000

                               Mn            Ni            Mo
1.5174E+03 BCC_A2            0.0003805     0.0013092     0.0000019
3.4154E+02 CEMENTITE         0.0169557     0.0003520     0.0000173
5.6465E+00 M23C6             0.0004359     0.0040889     0.0907447

                               Cr
1.5174E+03 BCC_A2            0.0005691
3.4154E+02 CEMENTITE         0.0771491
5.6465E+00 M23C6             0.1193824

Gibbs Energy = -5.0070853585E+07 J   System Enthalpy =  2.5323309115E+07 J
```

FIGURE 3.13   Example of calculated phase equilibria containing all the information that would be available from a phase diagram. At the temperature and pressure concerned, there are three phases in equilibrium, BCC_A2, CEMENTITE and M23C6 in a system containing 7 elements (Fe, C, Si, Mn, Ni, Mo, Cr).

3.6   IDEAL SOLUTION

An ideal solution is one in which the atoms at equilibrium are distributed randomly; the interchange of atoms within the solution causes no change in the potential energy of the system. For a binary (A-B) solution the numbers of the different kinds of bonds can therefore be calculated using simple probability theory:

$$N_{AA} = \frac{1}{2}zN(1-x)^2$$
$$N_{BB} = \frac{1}{2}zNx^2$$
$$N_{AB} = zN(1-x)x$$

where N_{AB} represents both A-B and B-A bonds which cannot be distinguished. N is the total number of atoms and x the fraction of B atoms. The factor of $\frac{1}{2}$ avoids counting A-A or B-B bonds twice. Here the term z is a coordination number.

For an ideal solution, the entropy of mixing is given by Equation 2.10. There is no enthalpy of mixing since there is no change in energy when bonds between like-atoms are broken to create those between unlike-atoms. This is why the atoms are randomly distributed in the solution. The molar free energy of mixing is therefore:

$$\Delta G_M = N_a kT [(1-x)\ln\{1-x\} + x\ln\{x\}]. \tag{3.8}$$

Figure 3.14 shows how the configurational entropy and the free energy of mixing vary as a function of the concentration. ΔG_M is at a minimum for the equiatomic alloy because that is when the entropy of mixing is at its largest; the curves are naturally symmetrical about $x = 0.5$. The form of the curve does not change with temperature though the magnitude at any concentration scales with the temperature.

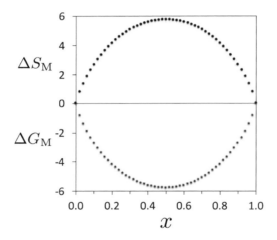

FIGURE 3.14 The entropy of mixing (J mol^{-1} K^{-1}) and the free energy of mixing (kJ mol^{-1}) as a function of concentration in an ideal binary solution where the atoms are distributed at random. The free energy is for a temperature of 1000 K. The data are plotted as dots rather than curves because concentration is strictly a discrete variable. So the slope at the vertical axes is not $\pm\infty$ as implied by Equation 3.8, but finite though very large.

3.7 REGULAR SOLUTIONS

There are no ideal solutions, though the heats of mixing in praseodymium-neodymium solid solutions come close to being ideal [29]. The iron-manganese liquid phase is also close to ideal, though even that has an enthalpy of mixing which is about -860 J mol^{-1} for an equiatomic solution at 1000 K, which compares with the configurational entropy contribution of about -5800 J mol^{-1}. The ideal solution model is nevertheless useful because it provides a reference. The free energy of mixing for a non-ideal solution often is written with an additional term, the *excess* free energy ($\Delta_e G = \Delta_e H - T\Delta_e S$) that accounts for the deviation from ideality:

$$\Delta G_M = \Delta_e G + N_a kT[(1-x)\ln\{1-x\} + x\ln\{x\}]$$
$$= \Delta_e H - T\Delta_e S + N_a kT[(1-x)\ln\{1-x\} + x\ln\{x\}] \qquad (3.9)$$

One of the components of the excess enthalpy of mixing comes from the change in the energy when new kinds of bonds are created during the formation of a solution. This enthalpy is, in the *regular solution* model, estimated from the pairwise interactions between adjacent atoms. The term *regular solution* was proposed by Hildebrand [30] to describe mixtures, the properties of which when plotted varied in an aesthetically regular manner; he went on to imply that a regular solution, although not ideal, would still contain a random distribution of the constituents. Following Guggenheim, the term regular solution is now restricted to cover mixtures that assume an ideal entropy of mixing but have a non-zero enthalpy of mixing [31].

In the regular solution model, the enthalpy of mixing is obtained by counting the different kinds of near neighbour bonds when the atoms are mixed at random; this information together with the binding energies gives the required change in the enthalpy on mixing. The binding energy may be defined by considering the change in energy as the distance between a pair of atoms is decreased from infinity to an equilibrium separation (Figure 3.15). The change in energy during this process is the binding energy, which for a pair of A atoms is written $-2\epsilon_{AA}$. It follows that when $\epsilon_{AA} + \epsilon_{BB} < 2\epsilon_{AB}$, the solution will have a larger than random probability of bonds between unlike atoms. The converse is true when $\epsilon_{AA} + \epsilon_{BB} > 2\epsilon_{AB}$ since atoms then prefer to be neighbours to their own kind. Notice that for an ideal solution it only is necessary for $\epsilon_{AA} + \epsilon_{BB} = 2\epsilon_{AB}$, and not $\epsilon_{AA} = \epsilon_{BB} = \epsilon_{AB}$.

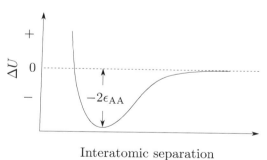

Interatomic separation

FIGURE 3.15 Change in energy as a function of the distance between a pair of A atoms. The term $-2\epsilon_{AA}$ is the binding energy for the pair of atoms. There is a strong repulsion at close-range.

Suppose now that the approximation that atoms are randomly distributed is retained, even though the enthalpy of mixing is not zero. The number of A-A, A-B and B-B bonds in a mole of solution is then $\frac{1}{2}zN_a(1-x)^2$, $zN_a(1-x)x$ and

$\frac{1}{2}zN_a x^2$ respectively, where z is the co-ordination number. It follows that the molar enthalpy of mixing is given by:

$$\Delta H_M \simeq N_a z(1-x)x\omega \qquad \text{where} \qquad \omega = \epsilon_{AA} + \epsilon_{BB} - 2\epsilon_{AB}. \qquad (3.10)$$

The product $zN_a\omega$ is often called the regular solution parameter, which in practice will be temperature and composition dependent. A composition dependence also leads to an asymmetry in the enthalpy of mixing as a function of composition about $x = 0.5$. For the nearly ideal Fe-Mn liquid phase solution, the regular solution parameter is $-3950+0.489T$ J mol^{-1} if a slight composition dependence is neglected.

A positive ω favours the clustering of like atoms whereas when it is negative there is a tendency for the atoms to order. This second case is illustrated in Figure 3.16, where an ideal solution curve is presented for comparison. Like the ideal solution, the form of the curve for the case where $\Delta H_M < 0$ does not change with the temperature, but unlike the ideal solution, there is a free energy of mixing even at 0 K where the configurational entropy term ceases to make a contribution.

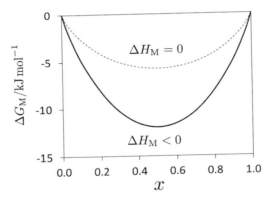

FIGURE 3.16 Free energy of mixing at 1000 K, as a function of concentration in a binary solution where there is a preference for unlike atoms to be near neighbours. The free energy curve for the ideal solution ($\Delta H_M = 0$) is included for reference.

The corresponding case for $\Delta H_M > 0$ is illustrated in Figure 3.17, where the form of the curve is seen to change with temperature. The contribution from the enthalpy term can largely be neglected at high temperatures where the atoms become randomly mixed by thermal agitation so the free energy curve then has a single minimum. However, as the temperature is reduced, the opposing contribution to the free energy from the enthalpy term introduces two minima at the solute-rich and solute-poor concentrations. This is because like-neighbours are preferred. On the other hand, there is a maximum at the equiatomic composition because that gives a large number of unfavoured unlike atom bonds. Between the minima and the maximum lie points of inflexion which are of importance in spinodal decomposition, to be discussed later.

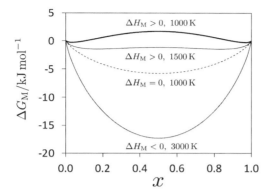

FIGURE 3.17 Free energy of mixing as a function of concentration and temperature in a binary solution where there is a tendency for like atoms to cluster. The free energy curve for the ideal solution ($\Delta H_M = 0$) is included for reference.

Chapter 4

Diffusion

Mass transport in a gas or liquid generally involves the flow of fluid (e.g., convection currents) although atoms also diffuse. Solids on the other hand, can support shear stresses and hence do not flow, except by diffusion involving the jumping of atoms on a fixed network of sites.

Consider first the random jumping of atoms in one dimension, with each jump through a distance that is unity; since the probability of a forward (+ve direction) or reverse (−ve direction) jump is equal, on completing a large number of steps, the total distance moved will be zero. Taking squares of these distances and then calculating the mean avoids this problem. The *mean* square of the distance moved during n jumps then becomes (the over-lines indicate mean values) [32]:

$$
\begin{aligned}
\overline{z^2} &= \overline{(z_1 + z_2 + z_3 \ldots z_n)^2} \\
&= \overline{(z_1 + z_2 + z_3 \ldots z_n)(z_1 + z_2 + z_3 \ldots z_n)} \\
&= (\overline{z_1^2} + \overline{z_2^2} + \ldots \overline{z_n^2}) + 2(\overline{z_1 z_2} + \ldots \overline{z_1 z_n} + \overline{z_2 z_3} \ldots \overline{z_2 z_n} + \ldots)
\end{aligned}
$$

$$(4.1)$$

where z_i is the ith step of length unity. Terms such as $z_1 z_2$ can have four possible values, (1×1), (1×-1), (-1×1) and (-1×-1), so their mean values will be zero. It follows that $\overline{z^2} = n$ so the root mean distance moved is $\sqrt{n}$.

Assume at first that such jumps can somehow be achieved in the solid state, with a frequency v with each jump over a distance λ.

For random jumps over a time period t, the mean distance is

$$\bar{z} = \lambda \sqrt{n} \qquad \text{where } n \text{ is the number of jumps} \qquad (4.2)$$
$$= \lambda \sqrt{vt} \qquad \text{where } t \text{ is the time}$$

diffusion distance $\propto \sqrt{t}$

Note that although the discussion is focused on atoms, the principles of diffusion apply to other species such as ions and molecules.

4.1 DIFFUSION IN A UNIFORM CONCENTRATION GRADIENT

Consider adjacent planes in an impure crystalline solid, as illustrated in Figure 4.1.

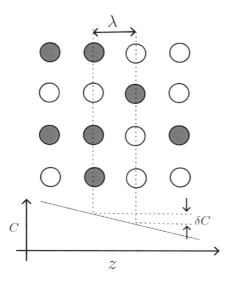

FIGURE 4.1 Diffusion gradient in a crystalline solid. λ is the shortest jump distance between adjacent planes. C is the solute concentration per unit volume and z is a distance.

The concentration of solute C, has units of number of atoms per unit volume (m^{-3}). Each plane therefore has $C\lambda$ atoms per unit area (m^{-2}) so the increment of composition on traversing a distance λ is

$$\delta C = \lambda \left\{ \frac{\partial C}{\partial z} \right\} \tag{4.3}$$

Given an atomic jump frequency v and six equally probable jump directions, it follows that the atom flux, J, atoms $\mathrm{m}^{-2}\,\mathrm{s}^{-1}$ in the forward (z) and reverse $(-z)$ directions is:

$$J_{\mathrm{L}\to\mathrm{R}} = \frac{1}{6} v C \lambda$$

$$J_{\mathrm{R}\to\mathrm{L}} = \frac{1}{6} v (C + \delta C) \lambda$$

Therefore, the net flux in the forward direction becomes

$$\begin{aligned} J = J_{\mathrm{L}\to\mathrm{R}} - J_{\mathrm{R}\to\mathrm{L}} &= -\frac{1}{6} v\, \delta C\, \lambda \\ &= -\frac{1}{6} v \lambda^2 \left\{ \frac{\partial C}{\partial z} \right\} \\ &\equiv -D \left\{ \frac{\partial C}{\partial z} \right\} \end{aligned} \tag{4.4}$$

This is Fick's first law where the constant of proportionality D is called the diffusion coefficient in $m^2\,s^{-1}$. Fick's first law applies to steady state flux in a uniform concentration gradient. Thus, the mean diffusion distance (Equation 4.2) can now be expressed in terms of the diffusivity as

$$\bar{z} = \lambda\sqrt{\nu t} \quad \text{with} \quad D = \frac{1}{6}\nu\lambda^2 \quad \text{giving} \quad \bar{z} = \sqrt{6Dt} \qquad (4.5)$$

Example 14: solidification of Al-Cu

An alloy of composition 90Al-5Cu wt% is slowly solidified. The microstructure obtained is illustrated in Figure 4.2, along with a phase diagram of the Al-Cu system.

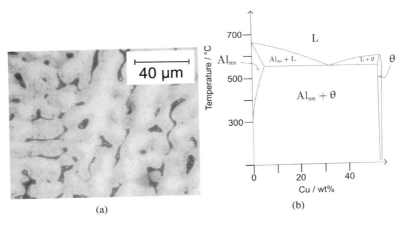

(a)

(b)

FIGURE 4.2 (a) Microstructure of solidified alloy. Image courtesy of T. W. Clyne under CC BY-NC-SA 2.0 UK license, DoITPoMS micrograph library [28]. (b) Part of the Al-Cu phase diagram.

(i) Sketch the important features of the microstructure illustrated, indicating the phases present.

(ii) Explain the development of the microstructure during cooling from the liquid to room temperature.

(iii) Estimate the fraction of the alloy expected to have the eutectic composition. How does this compare with the micrograph shown?

(iv) Give a rough estimate of the temperature needed to homogenise the microstructure in 800 s, given that the interdiffusion coefficient is $D = 5 \times 10^{-6}\exp\{-120,000/RT\}\,m^2\,s^{-1}$ with the gas constant $R = 8.3143\,J\,mol^{-1}\,K^{-1}$.

Solution 14

(i)   The light areas represent the primary Al-rich phase labelled Al_{ss} on the phase diagram, with the darker areas representing the eutectic mixture of Al_{ss} and θ.

(ii)   The primary Al_{ss} is expected to solidify at around 630 °C. The first solid to form is much less Cu-rich than the liquid ($\approx$ 1.5 wt%). As solidification proceeds, the liquid becomes progressively richer in Cu, but because diffusion may not occur at a rate consistent with equilibrium, so the solid is cored. When the eutectic temperature is reached the remaining liquid forms the eutectic.

    The primary Al_{ss} has developed as dendrites because of constitutional supercooling.

(iii)   The eutectic composition is about 32 wt% Cu, so using the lever rule, the mass fraction of eutectic expected under equilibrium conditions would be $(10 - 6)/(32 - 6) = 0.15$. This will not compare exactly with the micrograph, which may have a larger quantity of eutectic because of the fact that the Al_{ss} is cored.

(iv)   The coring is over a distance roughly 10 μm, so taking this as the diffusion distance z, and assuming random walk, $z = 6\sqrt{Dt}$, $D = z^2/(36t) = 3.47 \times 10^{-15}$ m^2 s^{-1}. Setting this equal to the diffusion coefficient, yields the required temperature as 684 K.

4.2   NON-UNIFORM CONCENTRATION GRADIENTS

Suppose that the concentration gradient is not uniform, Figure 4.3. Consider the region between planes 1 and 2.

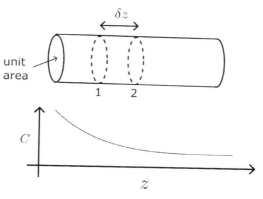

FIGURE 4.3   Non-uniform concentration gradient

$$\text{Flux in} \quad = -D\left\{\frac{\partial C}{\partial z}\right\}_1$$

$$\text{Flux out} \quad = -D\left\{\frac{\partial C}{\partial z}\right\}_2$$

$$= -D\left[\left\{\frac{\partial C}{\partial z}\right\}_1 + \delta z\left\{\frac{\partial^2 C}{\partial z^2}\right\}\right] \tag{4.6}$$

In the time interval δt, the concentration changes δC

$$\delta C \delta z = (\text{Flux in} - \text{Flux out})\, \delta t$$

$$\frac{\partial C}{\partial t} = D\frac{\partial^2 C}{\partial z^2} \tag{4.7}$$

assuming that the diffusivity is independent of the concentration. This is Fick's second law of diffusion.

4.2.1 Exponential solution

It is amenable to numerical solutions for the general case but there are a couple of interesting analytical solutions for particular boundary conditions. For a case where a fixed quantity of solute (C_B) is plated onto a semi-infinite bar, Figure 4.4,

$$\text{boundary conditions} \quad \begin{cases} \int_0^\infty C\{z,t\}\, \mathrm{d}z = B \\ C\{z, t = 0\} = 0 \end{cases}$$

the solution to Fick's 2nd law is

$$C\{z,t\} = \frac{B}{\sqrt{\pi D t}}\, \exp\left\{\frac{-z^2}{4Dt}\right\}. \tag{4.8}$$

4.2.2 Error function solution

Now imagine that we create the diffusion couple illustrated in Figure 4.5, by stacking an infinite set of thin sources on the end of one of the bars. Diffusion can thus be treated by taking a whole set of the exponential functions obtained above, each slightly displaced along the z axis, and summing (integrating) up their individual effects. The integral is in fact the error function

$$\text{erf}\{z\} = \frac{2}{\sqrt{\pi}}\int_0^z \exp\{-u^2\}\, \mathrm{d}u \tag{4.9}$$

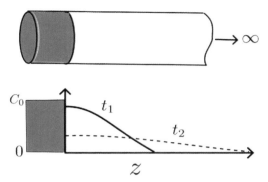

FIGURE 4.4 Exponential solution. Note how the curvature of the concentration distribution changes with time. The areas under the curves in the lower figure are identical.

so the solution to the diffusion equation subject to the following

$$\text{boundary conditions} \begin{cases} C\{z = 0, t\} = C_s \\ C\{z, t = 0\} = C_0 \end{cases}$$

is

$$C\{z, t\} = C_s - (C_s - C_0)\text{erf}\left\{\frac{z}{2\sqrt{Dt}}\right\} \qquad (4.10)$$

where C_0 is set to C_A or C_B depending on which side of the diffusion couple is considered.

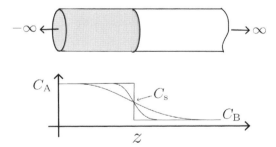

FIGURE 4.5 The error function solution. Notice that the "surface" concentration C_s remains fixed.

This solution can be used in many circumstances where the surface concentration is maintained constant, for example in the carburisation or decarburisation processes (the concentration profiles would be the same as in Figure 4.5, but with only one half of the couple illustrated). The solutions described here apply also to the diffusion of heat.

4.3 MECHANISM OF DIFFUSION

We have so far considered diffusion in a phenomenological manner, neglecting the details of the atomic mechanisms. Atoms in the solid-state migrate by jumping into vacancies, Figure 4.6. The vacancies may be interstitial or in substitutional sites. There is, nevertheless, a barrier to the motion of the atoms because the motion is associated with a transient distortion of the lattice.

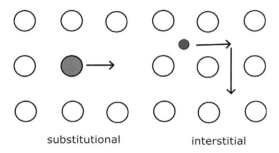

substitutional   interstitial

FIGURE 4.6   Mechanism of interstitial and substitutional diffusion. In the latter case, an atom next to the diffusing-atom is missing.

Assuming that the atom attempts jumps at a frequency ν_0, the frequency of successful jumps is given by

$$\nu = \nu_0 \exp\left\{-\frac{G^*}{kT}\right\}$$

$$\equiv \underbrace{\nu_0 \exp\left\{\frac{S^*}{k}\right\}}_{\text{independent of } T} \times \exp\left\{-\frac{H^*}{kT}\right\} \tag{4.11}$$

where k and T are the Boltzmann constant and the absolute temperature respectively, and H^* and S^* the activation enthalpy and activation entropy respectively. Since

$$D \propto \nu \qquad \text{we find} \qquad D = D_0 \exp\left\{-\frac{H^*}{kT}\right\} \tag{4.12}$$

A plot of the logarithm of D versus $1/T$ should therefore give a straight line, Figure 4.7, the slope of which is $-H^*/k$. Note that H^* is frequently called the activation energy for diffusion, designated Q.

The activation enthalpy of diffusion can be separated into two components, one the enthalpy of migration (due to distortions caused as the atom jumps) and the enthalpy of formation of a vacancy in an adjacent site. After all, for the atom to jump it is necessary to have a vacant site; the equilibrium concentration of vacancies can be very small in solids. Since there are many more interstitial

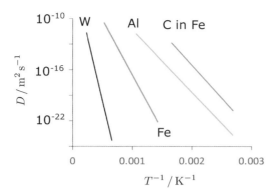

FIGURE 4.7 Typical self-diffusion coefficients for pure metals and for interstitial-carbon in ferritic iron. The uppermost diffusivity for each metal is at its melting temperature. Ferritic iron is body-centred cubic in crystal structure.

vacancies, and since most interstitial sites are vacant, interstitial atoms diffuse far more rapidly than substitutional solutes.

Example 15: interstitial diffusion

Given that $D_0 = 1 \times 10^{-6}$ m^2 s^{-1} and $Q = 84$ kJ mol^{-1} for the diffusion of carbon in iron, calculate the diffusivity at $1000\,^\circ$C.

A slice of pure iron, 1 mm thick, is placed in a carbon-containing atmosphere at a temperature of $1000\,^\circ$C, such that the carbon concentration C_s at the surface of the iron is maintained at 0.1 wt%. Calculate the concentration at a depth 0.1 mm below the surface after 1 min.

Bearing in mind that carbon penetrates the slice from both of the broad faces, sketch the concentration profile through the slice after 1 min at $1000\,^\circ$C.

Solution 15

Since $D = D_0 \exp\{-Q/RT\}$, the diffusion coefficient is 3.57×10^{-10} m^2 s^{-1}.

Use Equation 4.10 to show that $C\{z, t\} = C\{z = 0.1$ mm, 1 min$\} = 0.063$ wt%.

Carbon makes steel harder, but too much carbon also makes it brittle. So for some applications such as gears or aeroengine bearings, carbon is enriched at the surface by the diffusion process explained in the example above, while the

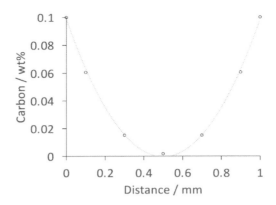

FIGURE 4.8 Calculate the concentrations at 0, 0.1 (already done above), 0.3 and 0.5 mm. Assume that the fluxes from the opposing faces do not interfere. You should get the profile illustrated.

core remains at a low-enough concentration to remain ductile. Figure 4.9 shows a micrograph form such a *carburised* steel, where the region at the surface looks darker because of the high carbon concentration there.

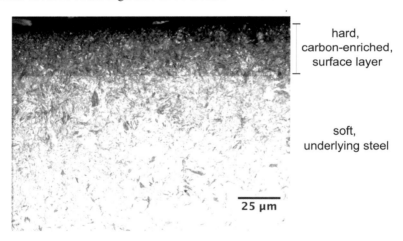

hard,
carbon-enriched,
surface layer

soft,
underlying steel

25 μm

FIGURE 4.9 A carburised steel. The image shows a cross-section through the surface, metallographically prepared and then etched with dilute nitric acid. The higher-carbon surface therefore looks darker in an optical microscope. Image adapted from one provided by Steve Ooi.

Chapter 5

Diffusion: other aspects

Diffusion in the solid state is at first sight difficult to appreciate. A number of mechanisms have been proposed historically, Figure 5.1. This includes a variety of ring mechanisms where atoms simply swap positions, but controversy remained because the strain energies associated with such swaps made the theories uncertain. One possibility is that diffusion occurs by atoms jumping into vacancies. But the equilibrium concentration of vacancies is typically 10^{-6}, which is small.

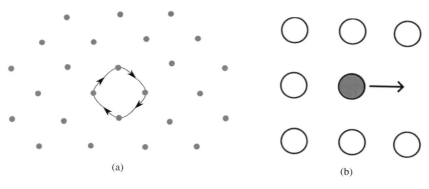

(a)   (b)

FIGURE 5.1 Schematic illustration of two of the possible solid-state diffusion mechanisms. a) Ring diffusion. (b) Vacancy-based diffusion.

The theory for the vacancy-based diffusion mechanism was therefore not generally accepted until an elegant experiment by Smigelskas and Kirkendall [33], Figure 5.2.

The experiment applies to solids as well as immiscible liquids. Consider a couple made from atoms 'A' and 'B'. If the diffusion fluxes of the two elements are different ($|J_A| > |J_B|$) then there will be a net flow of matter past the inert markers, causing the couple to shift bodily relative to the markers. This can only happen if diffusion is by a vacancy mechanism.

An observer located at the markers will see not only a change in concentration due to intrinsic diffusion, but also because of the Kirkendall flow of matter past the markers. The net effect is described by the usual Fick's laws, but with an interdiffusion coefficient $\overline{D}$ which is a weighted average of the two intrinsic

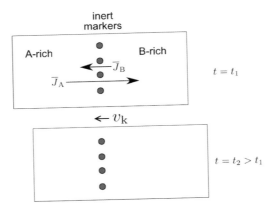

FIGURE 5.2 Diffusion couple with inert markers.

diffusion coefficients:

$$\overline{D} = x_B D_A + x_A D_B \tag{5.1}$$

where x represents a mole fraction. It is the interdiffusion coefficient that is measured in most experiments.

5.1 STRUCTURE-SENSITIVE DIFFUSION

Figure 5.3a shows the shape in three dimensions of space-filling grains in aluminium; the image is take using a scanning electron microscope from a neatly fractured piece of aluminium. We assume instead, the shape illustrated in Figure 5.3b in order to simplify the problem of the role of grain boundaries in enhancing the diffusion flux.

Crystals may contain non-equilibrium concentrations of defects such as vacancies, dislocations and grain boundaries. These may provide easy diffusion paths through an otherwise perfect structure. Thus, the grain boundary diffusion coefficient D_{gb} is expected to be much greater than the diffusion coefficient associated with the perfect structure, D_P.

Assume a cylindrical grain, Figure 5.3b. On a cross section, the area presented by a boundary is $2\pi r\delta$ where δ is the thickness of the boundary. Note that the boundary is shared between two adjacent grains so the thickness associated with one grain is $\frac{1}{2}\delta$. The ratio of the areas of grain boundary to grain is therefore

$$\text{ratio of areas} = \frac{1}{2} \times \frac{2\pi r\delta}{\pi r^2} = \frac{\delta}{r} = \frac{2\delta}{d} \tag{5.2}$$

where $d = 2r$ is the grain diameter.

For a unit area, the overall flux is the sum of that through the lattice and that

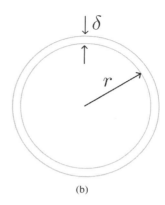

(a)   (b)

FIGURE 5.3 (a) Illustrating the three dimensional grain shapes in aluminium that has been fractured. Micrograph adapted from image courtesy of A. Shirzadi and E. Muyupa. (b) Grain shape
assumed for simplicity; δ is the thickness of the boundary.

through the boundary:

$$J = J_P\left(1 - \frac{2\delta}{d}\right) + J_{gb}\frac{2\delta}{d}$$

so that $\quad D_{measured} = D_P\left(1 - \frac{2\delta}{d}\right) + D_{gb}\frac{2\delta}{d}$ $\qquad$ (5.3)

Note that although diffusion through the boundary is much faster, the fraction
of the sample which is the grain boundary phase is small. Consequently, grain
boundary or defect diffusion in general is only of importance at low temperatures
where $D_P \ll D_{gb}$, Figure 5.4.

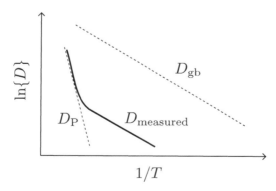

FIGURE 5.4 The actual diffusion coefficient will be due to the combination of fluxes through the
perfect regions of the crystal and through the grain boundaries.

Some solids, particularly polymers, relax under the influence of solutes which penetrate them. Alternatively, they react with the solute. For example, the structure of an assembly of polymeric molecules may change when penetrated by a solute, and indeed may undergo a change in volume. Some of these phenomena can be exploited in the design of delivery systems for medicines, where a slow release of chemicals may be advantageous in contrast to the immediate dissolution of tablets.

Example 16: diffusion through boundary and volume

(i) At a particular temperature, the coefficient D_P describing diffusion through the perfect lattice differs by three orders of magnitude relative to that describing diffusion along grain boundaries ($D_{gb} = 2 \times 10^{-11}\, m^2\, s^{-1}$) in a polycrystalline material. Stating any assumptions, calculate the net diffusion coefficient that represents both fluxes, given that the grain size of the material is $100\, \mu m$.

(ii) On the basis of your answer, giving reasons, comment on whether the diffusion data represent measurements made at a very high temperature, or not.

(iii) Why is it important to eliminate grain boundaries in a turbine blade that operates in the hottest part of the engine? Explain how the blade could be made into a single-crystal without any grain boundaries.

Solution 16

(i) Be aware that diffusion through the lattice is slower than through boundaries, therefore, the conclusion is reached that $D_P = 10^{-3} D_{gb}$. It is reasonable to assume that the grain boundary thickness $\delta = 1\, nm$. This means that the term $2\delta/d$ is small, so for the given grain size d,

$$D_{measured} \approx D_P + D_{gb}\frac{2\delta}{d}$$

$$= 10^{-3} \times (2 \times 10^{-11}) + (2 \times 10^{-11})\frac{10^{-10}}{100 \times 10^{-6}} \qquad (5.4)$$

$$= 2 \times 10^{-14}\, m^2\, s^{-1}$$

(ii) Since the net diffusion coefficient is essentially that same as D_P, the measurements correspond to high temperatures since the net area through which volume diffusion occurs is much greater than through the boundary, and at high temperatures this dominates the flux even though the activation energy for diffusion through the perfect lattice is larger.

(iii) Diffusion through grain boundaries, which is more rapid than through the perfect lattice, enhances the creep rate. A single crystal blade is made by solidifying the shape directionally, with a spiral at the bottom so that only the crystal growing fastest makes it through the spiral, leading to single-crystal solidification in the rest of the mould.

Example 17: turbine blades

What role do grain boundaries play in the performance of turbine blades used in the construction of aero-engines? How might their detrimental effects be mitigated?

Solution 17

Blades were originally polycrystalline, later to be replaced by directionally solidified blades, in which the grains are elongated along the growth direction, which is also the principal stress-axis in service. Since the type of creep concerned occurs by the elongation of individual grains via diffusion of vacancies along the length of the grain, this resulted in increased diffusion distances and reduced creep rates. Later, single crystal blades were produced, in which this type of creep was eliminated, since there were no grain boundaries. The three types of turbine blades that in some cases creep fast because of the presence of grain boundaries are illustrated in Figure 5.5.

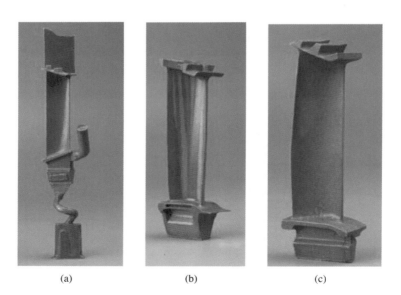

(a)   (b)   (c)

FIGURE 5.5 Turbine blades used in part of aeroengines where the temperature can exceed 1500 °C. (a) Single crystal blade (the spiral is later machined off), used to make the blade into a single crystal. (b) Columnar grains aligned to stress axis, made by solidifying in a temperature gradient. (c) Polycrystalline blade. The equiaxed grain structure is just visible.

5.2 THERMODYNAMICS OF DIFFUSION

Fick's first law is empirical in that it assumes a proportionality between the diffusion flux and the concentration gradient. However, diffusion occurs so as to minimise the free energy. It should therefore be driven by a gradient of free energy. But how do we represent the gradient in the free energy of a particular solute?

5.2.1 Chemical potential and diffusion

We first examine equilibrium for an allotropic transition (i.e., when the structure changes but not the composition). Two phases α and γ are said to be in equilibrium when they have equal free energies:

$$G^\alpha = G^\gamma \tag{5.5}$$

When temperature is a variable, the transition temperature is also fixed by the above equation, Figure 3.1.

A different approach is needed when chemical composition is also a variable (Section 3.1). Consider now the coexistence of two phases α and γ in our binary alloy. They will only be in equilibrium with each other if the A-atoms in γ have the same free energy as the A-atoms in α, and if the same is true for the B-atoms:

$$\mu_A^\alpha = \mu_A^\gamma \qquad \text{and} \qquad \mu_B^\alpha = \mu_B^\gamma$$

If the atoms of a particular species have the same free energy in both the phases, then there is no tendency for them to migrate, and the system will be in stable equilibrium if this condition applies to all species of atoms. Even then, we note that

$$x_A^{\alpha\gamma} \neq x_A^{\gamma\alpha} \qquad \text{and} \qquad x_B^{\alpha\gamma} \neq x_B^{\gamma\alpha}$$

where $x_i^{\alpha\gamma}$ is the mole fraction of i in phase α which is in equilibrium with γ etc. In spite of these differences in *concentration*, there is no tendency for diffusion.

5.2.2 Diffusion in a chemical potential gradient

It is proper therefore to say that diffusion is driven by gradients of free energy rather than chemical concentration:

$$J_A = -C_A M_A \frac{\partial \mu_A}{\partial z} \qquad \text{so that} \qquad D_A = C_A M_A \frac{\partial \mu_A}{\partial C_A} \tag{5.6}$$

where the proportionality constant M_A is known as the mobility of A. In this equation, the diffusion coefficient is related to the mobility by comparison with Fick's first law. Since μ_A is defined per unit concentration, it is multiplied by the concentration C_A in the J_A term.

The relationship is remarkable: if $\partial\mu_A/\partial C_A > 0$ as for the solution illustrated in Figure 3.4, then the diffusion coefficient is positive and the chemical potential gradient is along the same direction as the concentration gradient. However, if $\partial\mu_A/\partial C_A < 0$ then the diffusion will occur against a concentration gradient! This only can happen in a solution where the free energy curve has the form illustrated in Figure 3.17, repeated here for convenience in Figure 5.6.

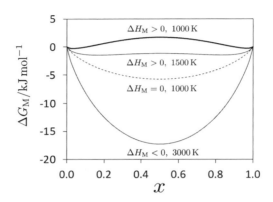

FIGURE 5.6 Free energy of mixing plotted as a function of temperature and of the enthalpy ΔH_M of mixing. $\Delta H_M = 0$ corresponds to an ideal solution where the atoms of different species always tend to mix at random and it is always the case that $\partial\mu_A/\partial C_A > 0$. When $\Delta H_M < 0$ the atoms prefer unlike neighbours and it is always the case that $\partial\mu_A/\partial C_A > 0$. When $\Delta H_M > 0$ the atoms prefer like neighbours so for low temperatures and for certain composition ranges $\partial\mu_A/\partial C_A < 0$ giving rise to the possibility of uphill diffusion.

Example 18: diffusion and thermodynamics

(i) Define what is meant by *uphill diffusion*. What sort of a free energy versus composition curve is necessary in order for uphill diffusion to be possible? What would be the signs of the enthalpy and entropy of mixing for a solution that exhibits uphill diffusion.

(ii) What is the condition that j-phases ($j > 1$) can coexist in equilibrium, if each contains i-components ($i > 1$)?

(iii) An alloy contains three different components such that its entropy of mixing is maximised. What is the concentration of each component. How could this entropy be increased?

Solution 18

(i) Uphill diffusion is when atoms diffuse up a concentration gradient, though the flow is down a chemical potential gradient. The curve of ΔG_M versus

x would have two minima at A-rich and B-rich locations, with a region in between where uphill diffusion is possible because a perturbation of composition would lead to a reduction in free energy. Such a solution would have a positive enthalpy of mixing and a positive entropy of mixing although $-T\Delta S_M$ would be negative.

(ii) Equilibrium between phases is defined by a uniform chemical potential for each component in all phases: $\mu_1^{\alpha} = \mu_1^{\beta} \ldots$ for all phases and components.

(iii) This is the maximum configurational entropy possible, which occurs when all components are in equal concentration. The maximum entropy of mixing can be increased by adding more components but ensuring that they all have equal concentrations.

Chapter 6

Nucleation

6.1 POSSIBLE MECHANISMS OF NUCLEATION

Phase fluctuations occur as random events due to the thermal vibration of atoms. An individual fluctuation may or may not be associated with a reduction in free energy, but it can only survive and grow if there is a reduction. There is a cost associated with the creation of a new phase, the interface energy, a penalty which becomes smaller as the particle surface to volume ratio decreases. In a metastable system this leads to a critical size of fluctuation beyond which growth is favoured.

Consider the homogeneous nucleation of α from γ. For a spherical particle of radius r with an isotropic interfacial energy $\sigma_{\alpha\gamma}$, the change in free energy as a function of radius is:

$$\Delta G = \frac{4}{3}\pi r^3 \Delta G_{\text{CHEM}} + \frac{4}{3}\pi r^3 \Delta G_{\text{STRAIN}} + 4\pi r^2 \sigma_{\alpha\gamma} \tag{6.1}$$

where $\Delta G_{\text{CHEM}} = G_V^\alpha - G_V^\gamma$, G_V is the Gibbs free energy per unit volume and ΔG_{STRAIN} is the strain energy per unit volume of α. The variation in ΔG with size is illustrated in Figure 6.1; the activation barrier and critical size obtained using Equation 6.1 are given by:

$$G^* = \frac{16\pi\sigma_{\alpha\gamma}^3}{3(\Delta G_{\text{CHEM}} + \Delta G_{\text{STRAIN}})^2}, \qquad r^* = -\frac{2\sigma_{\alpha\gamma}}{\Delta G_{\text{CHEM}} + \Delta G_{\text{STRAIN}}}. \tag{6.2}$$

The important outcome is that in classical nucleation the activation energy G^* varies inversely with the square of the driving force. Since the mechanism involves random phase fluctuations, it is questionable whether the model applies to cases where thermal activation is in short supply. In particular, G^* must be very small indeed if the transformation is to occur at a reasonable rate at low temperatures.

The nucleation rate per unit volume, I_V will depend on the frequency ν with which the embryo attempts to overcome the activation barrier; successful jumps depend on the probability $\exp\{-G^*/kT\}$ and the number of nucleation sites per unit volume (N_V):

$$I_V = \nu N_V \exp\left\{-\frac{G^*}{kT}\right\} \tag{6.3}$$

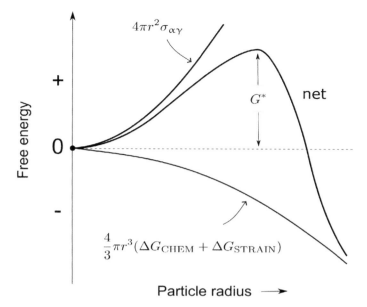

FIGURE 6.1 The activation energy barrier G^* and the critical nucleus size r^* according to classical nucleation theory based on heterophase fluctuations. It is assumed that the nucleus is spherical.

For homogeneous nucleation, each atom can be considered as a potential nucleation site, but it is more realistic to consider that a number of atoms together might form such a site. For the heterogeneous nucleation discussed below, N_V is dramatically reduced.

The nucleation barrier can be reduced if some defect present in the parent phase can be eliminated or altered during the process of nucleation; this phenomenon is known as *heterogeneous nucleation*. The barrier for heterogeneous nucleation is typically much smaller than that for homogeneous nucleation, and heterogeneities are so prevalent, that heterogeneous nucleation dominates in most practical scenarios. The defects responsible for heterogeneous nucleation are diverse: point defects, particles, dislocations, grain boundaries, surfaces etc. The characteristics of the heterogeneities are mostly insufficiently known to permit the quantitative prediction of nucleation rates. For this reason the theoretical focus is often on homogeneous nucleation, despite its limited practical relevance. Nevertheless, there are examples of quantitative, predictive analyses of heterogeneous nucleation. Two prominent cases are: the nucleation of freezing in metal castings, and the nucleation of the solid-state martensitic transformations.

Example 19: homogenous or heterogenous?

Figure 6.2 is a transmission electron micrograph showing the precipitation of extremely thin plates of vanadium carbide in iron. Explain whether the particles are homogeneously or heterogeneously nucleated? How could the number density of the particles be increased? Why might it be useful to increase the number density of precipitates?

Solution 19

The particles are heterogeneously nucleated on dislocations, i.e., line defects in the iron crystals. This is because some to the energy associated with these defects is relieved in the precipitation process, making them favourable sites. Whereas you can see that most particles are located on line defects, not all are. This might lead to the conclusion that some are homogeneously nucleated, but it would be a mistake to do so without detailed work. This is because in transmission electron microscopy, only very thin (100 nm) samples are electron-transparent, so there may be under- or over-lying regions where the defects are not visible.

Defects are incredibly difficult to avoid in real materials, so nucleation will then inevitably be heterogeneous; the gain in energy when a part of the defect is destroyed in the process of a nucleus forming, always works against homogeneous nucleation.

However, when one liquid nucleates in another, the process is homogeneous because the parent liquid has no structure and therefore no defects that can act as nucleation sites. Very hot oil and water (or liquid zinc and liquid lead) mix completely to form just one liquid. Cooling then leads to the homogeneous nucleation of droplets of oil in the water (or lead liquid in zinc liquid).

There are two ways of increasing the number density of particles. The first is to introduce more defects, for example, by subjecting the iron to plastic deformation before the heat treatment required to induce precipitation. The second is to reduce the activation energy by promoting precipitation at a temperature where the chemical driving force ΔG_{CHEM} is greater in *magnitude* (bear in mind that ΔG_{CHEM} is in fact negative for precipitation to occur at all). A larger magnitude of the chemical driving force means that the sample has been *supercooled* below the equilibrium temperature. Suppose that the equilibrium transformation temperature is T_e, then at equilibrium there is no driving force, $\Delta G_{CHEM} = 0$. If we write $\Delta G_{CHEM} = \Delta H - T\Delta S$ then at equilibrium, $\Delta H = T_e\Delta S$ and therefore, at any temperature,

$$\Delta G_{CHEM} = T_e\Delta S - T\Delta S = (T_e - T)\Delta S \qquad (6.4)$$

The term $(T_e - T)$ is known as the undercooling below the equilibrium temperature and assuming that ΔS does not vary much with temperature, is proportional to the chemical driving force ΔG_{CHEM}. The greater the undercooling the more is the parent phase supersaturated and wants to precipitate the product, but the temperature T must be reached sufficiently rapidly to ensure that transformation does not occur during cooling.

Now imagine the reverse transformation of the product into the parent during heating. The parent phase does not nucleate instantaneously when heating to T_e, so a *superheat* is required, in exact analogy to the discussion above.

Precipitates are obstacles to the progress of plastic deformation of the matrix. Therefore, if a greater strength is required, then N_V should be increased.

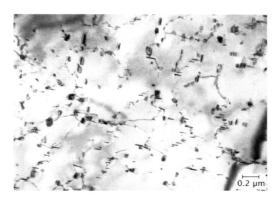

FIGURE 6.2   Precipitation of thin plates of vanadium carbide in iron.   Image courtesy of N. Ballinger.

6.2   MORE ABOUT HETEROGENEOUS NUCLEATION

Imagine that solid-α is nucleating from liquid (L) while in contact with an impenetrable solid surface (S), as illustrated in Figure 6.3.

The interfacial tensions at the junction between the phases must balance so

$$\sigma_{LS} = \sigma_{S\gamma} + \sigma_{L\gamma} \cos \theta \tag{6.5}$$

The surface area of the spherical cap is simply $2\pi r h \equiv 2\pi r^2 (1 - \cos \theta)$, and of the flat region between γ and the substrate, $\pi r^2 \sin^2 \theta \equiv \pi r^2 (1 - \cos^2 \theta)$, and finally, the volume V_n of the nucleus is:

$$V_n = \frac{\pi}{6} (3[r \sin \theta]^2 + h^2)h \equiv \frac{\pi r^3}{3} (2 - 3 \cos \theta + \cos^3 \theta). \tag{6.6}$$

Neglecting strain energy since the γ is surrounded by liquid, the net free energy change due to the formation of the spherical cap, bearing in mind that some L-S

69

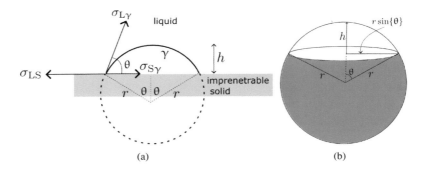

FIGURE 6.3 (a) Geometry of nucleation on a substrate, with the balancing of interfacial tensions. (b) Three-dimensional perspective.

interface is eliminated as the cap forms, L-γ and S-γ interface is created,

$$\Delta G = V_n \Delta G_{\text{CHEM}} + 2\pi r^2 (1 - \cos\theta)\sigma_{\text{L}\gamma} + \pi r^2 (1 - \cos^2\theta)(\sigma_{\text{S}\gamma} - \sigma_{\text{SL}}) \quad (6.7)$$

$$\therefore \frac{d\Delta G}{dr} = \pi r^2 (2 - 3\cos\theta + \cos^3\theta)\Delta G_{\text{CHEM}}$$

$$\underbrace{+ 4\pi r (1 - \cos\theta)\sigma_{\text{L}\gamma} + 2\pi r (1 - \cos^2\theta)(\sigma_{\text{S}\gamma} - \sigma_{\text{SL}})}_{\equiv 2\pi r\, \sigma_{\text{L}\gamma}\,(2 - 3\cos\theta + \cos^3\theta)} \quad (6.8)$$

with the equivalence indicated above coming from Equation 6.5, $\sigma_{\text{S}\gamma} - \sigma_{\text{SL}} = -\sigma_{\text{L}\gamma}\cos\theta$. It follows that critical size and activation energies during this form of heterogenous nucleation are given by

$$r^*_{\text{heterogenous}} = \frac{2\sigma_{\text{L}\gamma}}{|\Delta G_{\text{CHEM}}|} \quad (6.9)$$

$$G^*_{\text{heterogenous}} = \frac{4\pi}{3} \frac{\sigma_{\text{L}\gamma}^3}{|\Delta G_{\text{CHEM}}|^2} (2 - 3\cos\theta + \cos^3\theta). \quad (6.10)$$

Therefore, there is an advantage in heterogenous nucleation that some S-L interface is eliminated and replaced by S-γ so if $\sigma_{\text{LS}} > \sigma_{\text{S}\gamma}$, there is a gain of free energy in the process. As a result, the activation energy of heterogenous nucleation can be much smaller than that for the homogeneous scenario.

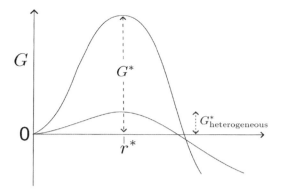

FIGURE 6.4 How the critical nucleus size and activation energy for nucleation changes during heterogenous nucleation.

Example 20: activation energy for heterogeneous nucleation

Prove the relationships given by equations 6.9 and 6.10.

Solution 20

Setting the differential in equation 6.8 to zero gives

$$0 = \pi r^2 (2 - 3\cos\theta + \cos^3\theta)\Delta G_{CHEM} + 4\pi r(1 - \cos\theta)\sigma_{L\gamma} + 2\pi r(1 - \cos^2\theta)(\sigma_{S\gamma} - \sigma_{LS})$$

Since

$$\sigma_{LS} = \sigma_{S\gamma} + \sigma_{L\gamma}\cos\theta = \sigma_{S\gamma} - \sigma_{LS} = -\sigma_\gamma\cos\theta,$$

it follows that

$$-\pi r^2(2 - 3\cos\theta + \cos^3\theta)\Delta G_{CHEM} = 2\pi r\sigma_{L\gamma}[2(1 - \cos\theta) + (1 - \cos^2\theta)(\cos\theta)]$$

which on rearrangement gives

$$r^*_{heterogenous} = \frac{2\sigma_{L\gamma}}{-\Delta G_{CHEM}} \equiv \frac{2\sigma_{L\gamma}}{|\Delta G_{CHEM}|} \tag{6.11}$$

The activation energy is obtained by substituting $r^*_{heterogenous}$ (written r^* here for brevity) into equation 6.7:

$$G^*_{\text{heterogenous}} = \frac{\pi(r^*)^3}{3}(2 - 3\cos\theta + \cos^3\theta)\Delta G_{\text{CHEM}} + 2\pi(r*)^2(1 - \cos\theta)\sigma_{L\gamma}$$

$$+ \pi(r^*)^2(1 - \cos^2\theta)(\sigma_{S\gamma} - \sigma_{LS})$$

$$= \pi(r^*)^2\left[\frac{1}{3}\frac{-2\sigma_{L\gamma}}{\Delta G_{\text{CHEM}}}\Delta G_{\text{CHEM}}(2 - 3\cos\theta + \cos^3\theta) + 2\sigma_{L\gamma}(1 - 2\cos\theta)\right.$$

$$\left. - \sigma_{L\gamma}\cos\theta + \sigma_{L\gamma}\cos^3\theta\right]$$

$$= \pi\left(\frac{-\sigma_{L\gamma}}{\Delta G_{\text{CHEM}}}\right)^2\frac{\sigma_{L\gamma}}{3}(2 - 3\cos\theta + \cos^3\theta)$$

$$= \frac{4\pi}{3}\frac{\sigma_{L\gamma}^3}{|\Delta G_{\text{CHEM}}|^2}(2 - 3\cos\theta + \cos^3\theta)$$

Notice that if the contact angle $\theta = 180°$ then the activation energy above becomes identical to that for homogeneous nucleation.

Example 21: nucleation

(i)   What would be the activation energy for nucleation in a system where the interfacial energy per unit area, between the parent and product phases, is zero?

(ii)   Why does heterogeneous nucleation have a smaller activation energy than homogeneous nucleation?

(iii)   Give three examples of locations in a polycrystalline material where heterogeneous nucleation would be favoured.

(iv)   What additional theory, apart from a calculation of the nucleation rate, would be needed to estimate the volume fraction of transformation as a function of time and temperature?

Solution 21

(i)   A zero interfacial energy would mean that nucleation is not necessary, the product phase can grow spontaneously once the equilibrium temperature is passed.

(ii)   Heterogeneous nucleation occurs on defects or inclusions. During heterogeneous nucleation, a part of the defect is destroyed, meaning that the free energy is reduced, making it favourable for nucleation to occur on defects.

(iii)   Grain boundaries of the parent phase, dislocations, free surfaces, or inclusions.

(iv) The growth rate of each nucleus would be important. The rest of the the answer requires thinking of concepts not discussed here. Particles growing from different nucleation sites will eventually impinge and such *hard* impingement would need to be accounted for. Similarly, the diffusion fields of particles could interfere before they touch, so this *soft* impingement needs to be accounted for.

6.3 NUCLEATING AGENTS

During solidification, it is possible to deliberately add particles that stimulate nucleation so that a smaller crystal size is obtained in the solid state. A small crystal size increases the strength and the ability of the material to absorb energy during impact with a foreign object.

Welding involves the deposition of molten metal within a gap between the components to be joined. Therefore, solidification begins at the junction between the solid and liquid and then proceeds towards the centre, resulting in columnar crystals, Figure 6.5a. In a particular application, such crystals are a disadvantage because they interfere with non-destructive testing of the final weld using ultrasonics. To avoid this, an inoculant designed to stimulate nucleation (TaN) in the liquid was added, resulting in an isotropic, uniform grain structure on solidification, Figure 6.5b.

(a)   (b)

FIGURE 6.5 Dramatic change in microstructure of a weld pool due to the stimulation of nucleation in the liquid. (a) Ordinary weld in a stainless steel. (b) The weld pool was inoculated with tantalum nitride particles. After Tyas and Bhadeshia.

High-speed steels are important as machining tools because they have outstanding wear and heat resistance. Figure 6.6 shows how the structure of such steels can be improved by inoculating the melt with tungsten carbide particles.

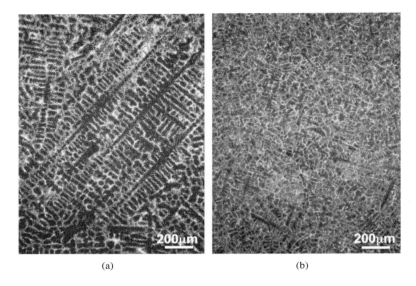

(a)   (b)

FIGURE 6.6   Dramatic change in microstructure of a high-speed steel due to the stimulation of nucleation in the liquid. (a) Ordinary microstructure after solidification. (b) The liquid pool was inoculated with tungsten carbide particles.   Images courtesy of Alexander Chaus of the Slovak University of Technology.

There are many conditions that must be satisfied for a material to be a good inoculant, the foremost of which it that it should have a higher melting temperature than the liquid to be inoculated. Good inoculants work either by lattice matching or by a chemical reaction which is locally favourable to the formation of the solid.

Sometimes, the first solid to form in a melt has a low strength because of the high temperatures involved. Any turbulence in the liquid can then cause some fragmentation. It is these fragments which may then grow into full-sized crystals. However, this is not strictly a nucleation event because the fragments usually are larger than a critical nucleus size.

Chapter 7

Solidification

We have seen that homogeneous nucleation is difficult. For example, in container-less experiments where a pure liquid is isolated from its environment by levitation or by allowing it to free fall over a long distance, it is possible to supercool the liquid well below it equilibrium freezing temperature ($\approx 200\,°C$). In general, however, solidification proceeds from the surfaces of the container (the mould) into which the liquid is poured, by heterogeneous nucleation at the container surface in the manner described on page 69.

The velocity v of the transformation front is related to the difference in the rate of liquid→solid atom jumps and the solid→liquid atom jumps. It is seen from Figure 7.1 that this gives:

$$v \propto \underbrace{\exp\left\{-\frac{Q}{RT}\right\}}_{\text{liquid}\rightarrow\text{solid}} - \underbrace{\exp\left\{-\frac{Q+|\Delta G|}{RT}\right\}}_{\text{solid}\rightarrow\text{liquid}} \qquad (7.1)$$

It follows that for small $|\Delta G|$, $v \propto |\Delta G|$ and to the undercooling $\Delta T = T_m - T$ below the equilibrium melting temperature T_m with $v \propto \Delta T$. Note that these proportionalities are based on the fact that for small x, $\exp\{x\} \simeq 1 + x$.

Figure 7.2 shows schematically the grain structures possible. The chill zone contains fine crystals nucleated at the mould surface, rapidly as the hot liquid is poured in to contact a cold mould surface. There is then selective growth into the liquid as heat is extracted from the mould. The growth occurs opposite to the direction of heat flow. If the liquid in the centre of the mould is undercooled sufficiently there may also be some equiaxed grains forming, independently of the mould surface.

Figure 7.2c illustrates thermal dendrites of ice forming on the internal surface of a window with the temperature outside being $-20\,°C$. As explained in the figure caption, a negative temperature gradient develops that leads to interfacial instability and hence, dendritic solidification of the condensed water. This is because a small perturbation at the interface ends up in even more supercooled liquid so the interface becomes unstable.

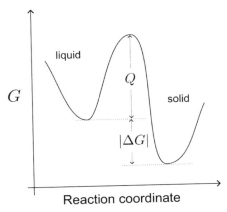

FIGURE 7.1 Barrier to solid-liquid interface motion. The transfer of atoms from the liquid to the solid will be faster than the reverse process, which has a much greater activation barrier $Q + |\Delta G|$.

Dendrites in metals have preferred growth directions, almost always $\langle 1\ 0\ 0 \rangle$ for those that have a cubic crystal structure. Lithium has a body-centred cubic crystal structure and dendrite growth along $\langle 1\ 0\ 0 \rangle$. During the recharging of lithium-ion batteries, if the lithium deposits too rapidly at the negative electrode (due for example to rapid charging) lithium dendrites form that can lead to a short circuit, overheating and sometimes cause the battery to burn.

Figure 7.3 shows a photograph of an aluminium ingot (about 5 cm wide) made by casting commercially pure aluminium into a metallic mould. The liquid in contact with the mould wall solidifies relatively rapidly, giving a fine, equiaxed grain structure there. This zone is called the *chill zone*. The grains then start to grow towards the centre, opposite the general direction of maximum heat flow, giving a columnar-grain structure. Those grains which have their maximum growth directions most antiparallel to the heat flow stifle the growth of less favourably oriented grains, leading to a coarsening of the columnar structure as solidification progresses towards the centre. There is sometimes a region of equiaxed grains in the centre of the ingot, arising from solid particles that form at the liquid surface, or from fragments of metal solidifying from the mould wall. The central pipe is due to the contraction of the liquid as it solidifies.

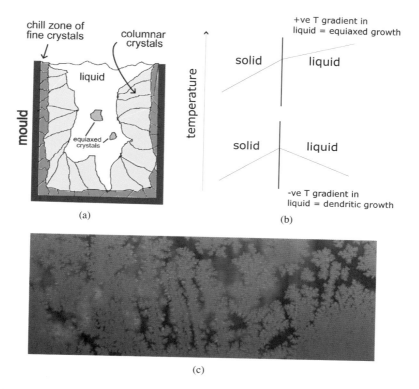

(a)

(b)

(c)

FIGURE 7.2 (a) Schematic representation of structure evolution when hot liquid is poured into a cold mould. (b) The effect of the temperature gradient in the liquid, on the evolution of grain structure during solidification. (c) Dendrites of ice forming on the inner surface of a building-window, the temperature outside being $-20\,^{\circ}C$. The temperature where moisture condenses is greater than the moisture-free glass, so there is a negative temperature gradient that encourages interface-instability, hence the ice dendrites. Image taken during a visit to the Harbin Institute of Technology, which is located in a region where winter temperatures are so low, that the region holds a major ice-sculpture festival each year (https://www.phase-trans.msm.cam.ac.uk/2004/HIT3/HIT3.html). More photographs of dendrites on https://www.phase-trans.msm.cam.ac.uk/dendrites.html

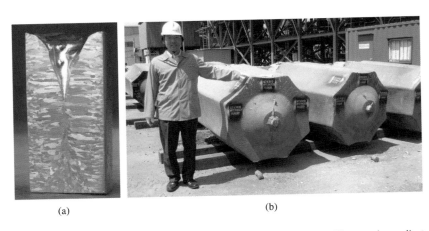

(a)

(b)

FIGURE 7.3 (a) Longitudinal section through a small aluminium casting to illustrate the gradients of microstructure. (b) Much larger ingots of stainless steel manufactured at the POSCO factory in South Korea. More images available on https://www.phase-trans.msm.cam.ac.uk/2010/ingots.html.

7.1 ALLOYS: SOLUTE PARTITIONING

Dendrite formation is extremely common in alloys, where solute partitions between the solid and liquid phases, as indicated by the phase diagram in Figure 7.4. By convention, we shall label the composition of the solid phase which is in equilibrium with the liquid as C^{SL} and a similar interpretation applies to C^{LS}. C_0 represents the average composition of the alloy. The phase boundary defining C^{SL} as a function of temperature is known as the solidus because it separates the completely solid phase field from the mixture of solid and liquid. Similarly, that boundary defining C^{LS} is known as the liquidus, below which solid can precipitate.

The partition coefficient k_p is written

$$k_p = \frac{C^{SL}}{C^{LS}} \qquad \text{frequently} < 1$$

Under equilibrium conditions the compositions of the solid and liquid at all stages of solidification are given by a tie-line of the phase diagram, and the proportions of the phases at any temperature by the lever rule.

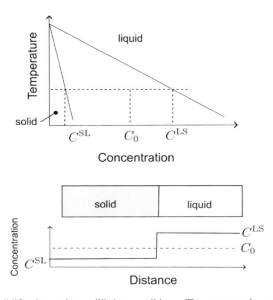

FIGURE 7.4 Solidification under equilibrium conditions. The concentration profile in the lower diagram shows that the solid and liquid compositions at the temperature considered are given by equilibrium, with no gradients of composition either in the solid or in the liquid.

7.2 NON-EQUILIBRIUM SOLIDIFICATION

In practice, the cooling conditions required to maintain equilibrium at all temperatures are too slow. The rate will be such that equilibrium is maintained only locally at the interface. Assuming that diffusion in the solid is too slow to allow its composition to homogenise as C^{SL} changes, a gradient of composition develops in the solid phase.

Referring to Figure 7.5, we note that

- solidification begins at a small undercooling below the melting temperature T_{m}, with the formation of a small amount of solid of composition C_0^{SL} which is much less than the composition of the liquid that is close to C_0.
- The equilibrium compositions obviously change with temperature according to the phase diagram; thus, both C^{SL} and C^{LS} increase as the temperature is reduced. Since diffusion is difficult in the solid phase, the first solid to form will therefore have a different chemical composition to subsequent solid. On reaching a temperature T_1, this lack of diffusion in the solid leads to a rise in its concentration to C_1^{SL} while the part that solidified early remains at C_0^{SL}.
- On reaching the temperature T_2, the alloy should be completely solidified under equilibrium conditions but is not. This is because solute is not uniformly distribute in the solid phase. However, the partitioning of solute between the solid and liquid ceases because the composition of the solid is now C_0. Therefore, the final liquid that will solidify at $T \ll T_2$ will contain all the solute partitioned during the cooling process.
- The microstructure of the finally solidified alloy will reveal concentration gradients. This is because for metallography, the sample is first polished flat and then attacked with an etchant, which reacts differently to regions of varying composition. An example is illustrated in Figure 7.6 but this kind of micro-segregation is common over a vast range of materials because it simply is impractical to solidify under equilibrium conditions.

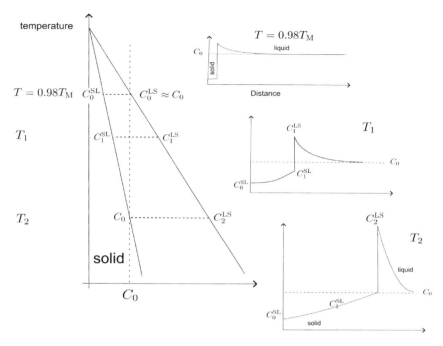

FIGURE 7.5 Solidification during the cooling of a liquid alloy under nonequilibrium conditions, when concentration gradients develop in the solid because diffusion there is slow. On the left is the equilibrium phase diagram and on the right, a variety of concentration profiles that develop at the temperatures indicated.

FIGURE 7.6 The gentle variation in contrast is due to the uneven distribution of manganese in solid solution within Fe-0.36C-0.58Mn-0.27Si wt% solidified at about 0.1 °C s^{-1} into a dendritic microstructure. The particularly dark regions represent the last liquid to solidify and have a manganese concentration of about 1.6 wt%. Image from *Guide to the Solidification of Steels*, courtesy of Elisabeth Nilsson, President of Jernkontoret, Sweden.

7.3 CONSTITUTIONAL SUPERCOOLING

Solute is partitioned into the liquid ahead of the solidification front. This causes a corresponding variation in the liquidus temperature, below which solid can precipitate. There is more often than not, a positive temperature gradient in the liquid, giving rise to a supercooled zone of liquid ahead of the interface, Figure 7.7. This is called *constitutional supercooling* [34] because it is caused by composition changes within the liquid in the presence of a positive temperature gradient ahead of the solidification front.

A small perturbation on the interface will therefore expand into a supercooled liquid, resulting in dendritic solidification.

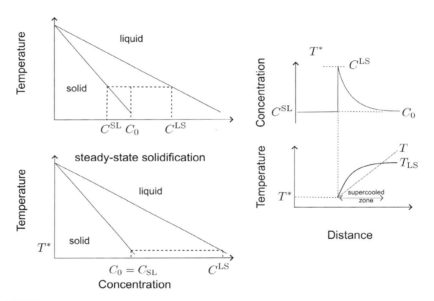

FIGURE 7.7 Diagram illustrating constitutional supercooling. T^* represents a temperature where solidification occurs at a steady pace, because the solid composition C_{SL} equals the average C_0, so no further solute is partitioned into the liquid. The composition profiles on the right therefore represent stead-state solidification. There is a region ahead of the solid-liquid interface in which the liquid is supercooled because the partitioned solute affects the liquidus temperature.

A supercooled zone of this kind occurs only when the liquidus temperature (T_{LS}) gradient at the interface is greater than the actual temperature gradient:

$$\left.\frac{\partial T_{LS}}{\partial z}\right|_{z=0} > \frac{\partial T}{\partial z} \quad \text{i.e.,} \quad \left.m\frac{\partial C_{LS}}{\partial z}\right|_{z=0} > \frac{\partial T}{\partial z}$$

where $m = \partial T_{LS}/\partial C_{LS}$ is the slope of the liquidus phase boundary on the phase diagram and the coordinate z has its origin at the solid-liquid interface.

The effective solute diffusion-distance is D/v (*cf.* Equation 1.1) where v is the velocity of the solid-liquid interface; therefore, the gradient of concentration at the interface (position $z = 0$) is

$$\left.\frac{\partial C_{LS}}{\partial z}\right|_{z=0} \approx -\frac{C^{LS} - C^{SL}}{D/v}$$

so that the minimum thermal gradient required for a stable solidification front is

$$\frac{\partial T}{\partial z} = \frac{mvC_0}{D}\left(\frac{1}{k_p} - 1\right) = \frac{mC_0(1 - k_p)v}{k_pD} \tag{7.2}$$

where the partition ratio $C^{SL} = k_pC^{LS}$. It is difficult to avoid constitutional supercooling in practice because the velocity required is very small indeed. Directional solidification with a planar front is possible only at slow growth rates, for example in the production of silicon single crystals for electronic devices. In most cases the interface is unstable so dendrites are frequently observed to form.

Example 22: solidification, interface stability

(i) Explain the circumstances in which a pure liquid might solidify with the growth of dendrites rather than by the movement of a planar liquid-solid interface.

(ii) Why is it possible to obtain dendritic solidification of an alloy when the temperature increases as a function of distance ahead of the solidification front?

(iii) Which of these would reduce the tendency to form dendrites?

- Large solute concentration.
- Low growth rate.
- Large, positive temperature gradient in the liquid ahead of the solidification front.
- Small partition coefficient.

Solution 22

(i) If the temperature gradient ahead of the solidification front is positive (T increasing with z) then any perturbation on the interface will decay because it advances into hotter liquid. So the gradient must be negative for the dendritic solidification of a pure liquid.

(ii) In the case of an alloy, it is the liquidus temperature at any location that must be compared against the temperature. Because of solute partitioning, the liquidus temperature may fall below the actual temperature of the liquid (even when the temperature gradient is positive), leading to a *constitutionally supercooled zone* ahead of the interface.

(iii) A large solute concentration is not particularly relevant because it is the partitioning of solute that matters.

A slow growth rate increases the diffusion distance D/v, thus levelling out the variation in concentration ahead of the interface. Therefore, the liquidus temperature gradient may become less than the actual temperature gradient, thus eliminating dendritic solidification.

A large positive temperature gradient would also reduce or eliminate the constitutionally supercooled zone, meaning that dendrite formation becomes difficult or impossible.

A large partition coefficient would help because it may reduce the change in liquidus temperature as a function of distance ahead of the interface.

Chapter 8

Iron-carbon system

Alloys of iron are, with the exception of concrete, by far the most successful of structural materials; there are simply no challengers for the vast majority of applications. Just one interesting example – Elon Musk's largest rocket, the *Starship* is made from steel, not aluminium, nor carbon fibre composites.

The vast majority of steels (Fe, C) are in the austenitic condition at temperatures in excess of 900 °C. Austenite has a cubic-close packed crystal structure and tends to decompose into ferrite (body-centered cubic) and cementite (Fe_3C, orthorhombic lattice), Figure 8.1.

Steels with a carbon concentration less than 0.76 wt% are hypoeutectoid, those with greater concentrations are hypereutectoid. A steel with exactly 0.76 wt% carbon will tend to decompose into an intimate mixture of cementite and ferrite at 723 °C by a eutectoid reaction, Figure 8.1:

$$\gamma \rightarrow \alpha + Fe_3C \qquad (8.1)$$

This mixture is known as *pearlite* where the ferrite and Fe_3C grow together at a common transformation front with the parent austenite, such that in Fe-C alloys, the average composition of the pearlite is the same as that of the austenite.

Example 23: pearlite

Why does pearlite in an Fe-C alloy grow at a constant rate.

Solution 23

Pearlite consists of α and Fe_3C the former being carbon-depleted and the latter, carbon-rich, when compared with the composition of the parent austenite from which they both grow simultaneously at a common transformation front. The ferrite partitions carbon into the adjacent austenite but the cementite absorbs the partitioned carbon as it grows, leaving the chemical composition of the bulk of the austenite unchanged. In other words, the average composition of the pearlite is identical to that of the austenite. Therefore, since nothing changes in the austenite, the pearlite grows at a constant rate.

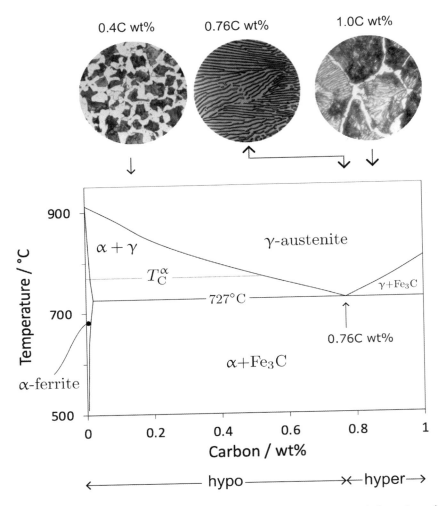

FIGURE 8.1 Iron-carbon equilibrium phase diagram. Austenite and ferrite are designated γ and α respectively. The ferrite changes from a paramagnetic to a ferromagnetic state on cooling below the Curie point T_C^α. The eutectoid reaction at 727 °C, 0.76C wt% is when austenite decomposes simultaneously into ferrite and cementite, according to Equation 8.1. At the eutectoid temperature, the solubility of carbon in ferrite that is in equilibrium with cementite is just 0.02 wt%. The micrographs represent a hypoeutectoid steel in which ferrite forms first followed by pearlite, a fully pearlitic eutectoid steel and a hypereutectoid steel that has layers of proeutectoid cementite and pearlite.

The diameters of the circular images are 400 µm, 40 µm and 40 µm. The images are adapted from the DoITPoMS micrograph library under CC BY-NC-SA 2.0 UK license [28] with the exception of the eutectoid steel, which is from the book "The alloying elements in steel" [35], that can be downloaded from https://www.phase-trans.msm.cam.ac.uk/2004/Bain.Alloying/ecbain.html

A hypoeutectoid steel will first decompose during cooling to ferrite at the austenite grain boundaries; the austenite is consequently enriched in carbon and eventually, when its composition reaches the eutectoid concentration, it decomposes into pearlite. The final microstructure will therefore be a mixture of ferrite and pearlite.

Example 24: fraction of cementite in pearlite

A eutectoid steel has the composition Fe-0.76C wt%. Cementite (Fe_3C) contains 6.67 wt% of carbon. Calculate the mass fraction of cementite in the pearlite that forms when the steel is cooled through the eutectoid temperature.

Solution 24

The solubility of carbon in ferrite that is in equilibrium with cementite is 0.02 wt% at the eutectoid temperature (Figure 8.1). Therefore, using the lever rule (Equation 3.6), the mass fraction of cementite is

$$\frac{0.76 - 0.02}{6.67 - 0.02} = 0.11$$

Note that the equilibrium fraction will increase slightly as the temperature is reduced because the solubility of carbon in ferrite that is in equilibrium with cementite decreases sharply with temperature.

8.1 MARTENSITE IN STEELS

Martensite is not an equilibrium phase. When austenite is cooled sufficiently rapidly, it transforms without diffusion into martensite, so there is no change in chemical composition. There is a martensite-start temperature or M_S below which it forms, representing the temperature at which the free energy change $\Delta G^{\gamma \to \alpha'}$ becomes negative (i.e., a reduction in free energy when γ-austenite transforms into α'-martensite). Martensite is not the same as ferrite in many respects:

- its chemical composition is identical to the of the parent γ-phase, so it is not represented on the equilibrium phase diagram Figure 8.1.
- There will be a shape deformation accompanying the transformation of austenite to martensite, of the type illustrated in Figure 1.3a. This results in strain energy. The shape change also causes the formation of defects, that add to the stored energy of martensite.
- Because of these departures from equilibrium, greater free energy change is required to stimulate martensite when compared against that needed for

equilibrium ferrite to precipitate: $|\Delta G^{\gamma \to \alpha'}| \gg |\Delta G^{\gamma \to \alpha}|$, where the vertical lines imply magnitudes. This is true even when the martensitic transformation of pure iron is considered, because the strain energy term still features.

The shape deformation that accompanies the formation of martensite introduces a lot of strains in the surrounding matrix. To minimise this strain energy, it grows in the form of thin plates, Figure 8.2.

FIGURE 8.2   Optical micrograph illustrating plates of martensite embedded in austenite, in a steel containing 1 wt% carbon, quenched from its fully austenitic state.   Image courtesy of Saurabh Chatterjee [36].

Example 25: calculation of M_S

The driving force for the transformation of austenite without a change in chemical composition varies with temperature as follows:

$$\Delta G^{\gamma \to \alpha} = 1.2T - 1000 \text{ J mol}^{-1}$$

where T is the absolute temperature. Calculate the martensite-start temperature if the stored energy of martensite is 700 J mol^{-1}.

Solution 25

Set $\Delta G^{\gamma \to \alpha}$ to equal the negative value of the stored energy, because that is how much the driving force must supply before martensite is triggered, and set the temperature to the unknown M_S:

$$\therefore -700 = 1.2 M_S - 1000 \qquad \text{so that} \qquad M_S = 250 \text{ K}$$

The fraction of martensite increases with the undercooling below M_S.

The martensite in steels is supersaturated with carbon. Carbon occupies octahedral interstices in the b.c.c. lattice; these are characterised by three principal axes $a\langle 0\,0\,1\rangle$, $a\langle 1\,1\,0\rangle$ and $a\langle 1\,\bar{1}\,0\rangle$, where a is the lattice parameter. There are three sub-lattices of octahedral holes, along directions parallel to the unit cell edges, Figure 8.3.

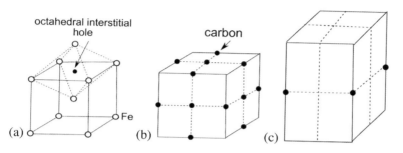

FIGURE 8.3   (a) The octahedral holes between the iron atoms which carbon can occupy. (b) Carbon cannot occupy all the sites, but here we assume that the carbon is randomly distributed so that the ferrite lattice remains cubic. (c) The ferrite lattice becomes tetragonal (greatly exaggerated for illustration purposes) if the carbon atoms occupy sites along just one edge of the unit cell.

Each carbon atom causes a tetragonal distortion since the principal axes of the octahedral sites are not equivalent. There is an expansion along $a\langle 1\,0\,0\rangle$ and small contractions along the other two axes. As a consequence, it is favourable for all the carbon atoms to lie on a single sub-lattice of octahedral interstices, giving rise to a body-centered tetragonal structure for the martensite, Figure 8.3c.

Each carbon atom acts as a strain centre. Because this strain field is tetragonal, it is particularly effective in interfering with dislocation motion since it interacts with both the shear and dilatational components of the stress field of dislocations. This is why carbon hardens martensite much more than it hardens austenite (where the octahedral hole is symmetrical, with axes along $a\langle 1\,0\,0\rangle$). Carbon-free martensite is not particularly strong.

8.2   TEMPERING OF MARTENSITE

Martensite containing carbon is very strong; this often makes it brittle. To achieve a compromise between strength and toughness, the martensite is tempered, i.e., heat treated at temperatures below that at which austenite can form. The heat treatment causes the following changes:

1. 200-400 °C. The precipitation of excess carbon at first as a transition carbide $Fe_{2.4}C$, which then converts to cementite (Fe_3C) as equilibrium is approached. This is accompanied by a significant loss of strength but an improvement in

toughness, Figure 8.4.

2. 400-500 °C. Recovery with a reduction in dislocation density. Cementite begins to coarsen with further loss in strength.

3. > 500 °C. Recrystallisation of plates into equiaxed grains of ferrite.

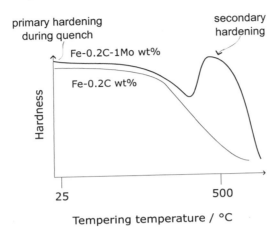

FIGURE 8.4 Martensite in steels is supersaturated with carbon. Tempering it for 1 h at the temperatures indicated allows the carbon to precipitate, leading to a dramatic drop on hardness for the Fe-C. However, in Fe-C-Mo, at temperatures where Mo becomes mobile, Mo_2C precipitates and leads to a late increase in hardness, i.e., secondary hardening.

Some steels contain strong carbide forming elements such as V, Mo, Cr or W. These are in substitutional solid solution and immobile unless the tempering temperature is in excess of about 500 °C. When they precipitate to form carbides such as VC, Mo_2C, $Cr_{23}C_6$ or W_2C, there is an increase in hardening, Figure 8.4, i.e., secondary hardening. This is because the carbides are very fine and frequently have coherency strain fields. Microstructures generated by secondary hardening are very stable and form the bulk of the alloys used in the power plant industry where the steam temperature is typically 500-620 °C and service life typically 40 years.

8.3 TIME-TEMPERATURE-TRANSFORMATION (TTT) DIAGRAMS

Martensitic transformations are not equilibrium and cannot therefore be represented on an equilibrium phase diagram. However, the M_S temperature can be drawn as a horizontal line on a temperature-versus time plot. If a sample is cooled to a particular temperature and held isothermally, there will be no martensite unless the temperature is below M_S.

Similarly, in a eutectoid steel, pearlite will not form unless austenite is cooled to below the eutectoid temperature. The pearlite transformation involves diffusion and hence will be slow at low temperatures. It will also be slow close to the eutectoid temperature. This can be represented on the $T - t$ plot as a C-shaped curve, Figure 8.5. The diagram is constructed by rapidly cooling austenite to the specified temperature and measuring the fraction of transformation as a function of time.

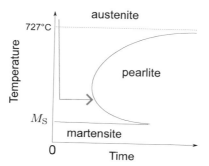

FIGURE 8.5 TTT diagram for eutectoid steel. The red arrow show how austenite is supercooled to the temperature corresponding to the horizontal arrow, and the time taken for transformation to pearlite to being.

A real TTT diagram is somewhat more complicated with many other austenite-transformation products possible and the overall kinetics are influenced by alloying elements other than carbon, Figure 8.6, which illustrates the circumstances for a hypoeutectoid steel. The effect of manganese on slowing down the transformation of austenite is also illustrated, important because we sometimes deal with large components which require time to reach the isothermal transformation temperature.

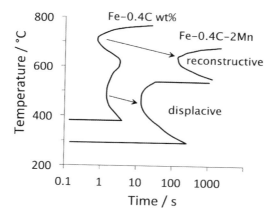

FIGURE 8.6 TTT diagrams for two hypoeutectoid steels, one containing manganese which retards all of the transformations from austenite.

In fact, with large components such as the rotors used to make steam turbines,

Figure 8.7, it would not be possible to achieve isothermal transformation because the surface would cool faster than the interior. So the alloy must be designed such that the rate of transformation ensures the correct microstructure at ambient temperatures. But the design of alloys for specific purposes is a big and active subject that has to be reserved for advanced studies.

FIGURE 8.7 A partial image of a steam turbine that is used in equipment for the generation of electricity.

Chapter 9

Miscellaneous

9.1 ALUMINIUM-COPPER ALLOYS

Some of the aluminium alloys that are used in making airframes contain copper which is added for strength achieved by precipitation hardening. Copper can be added in concentrations up to about 10 wt% although alloys for structural applications usually contain smaller concentrations, of the order of 4 wt%. Copper has a limited solubility in aluminium at low temperatures. An alloy that is quenched from high temperatures to retain the copper in solid solution will therefore be metastable. Given an opportunity the copper will tend to precipitate. This can occur even at room temperature, so that hardness will change as a function of time, a phenomenon known as *age hardening*.

The shape of the $Al_{ss}/Al_{ss} + \theta$ phase boundary is ideal for precipitation hardening because the solubility of the solute is large at high temperatures, so the sample can be cooled rapidly to a lower temperature and held there in order to precipitate the excess solute. There are many systems in metallurgy that are blessed with such a phase boundary because at high temperatures the free energy terms have greater contributions from entropy (i.e., $-T\Delta S$ term) and thus favour mixing whereas lower temperatures do not.

Aluminium itself, unlike iron or plutonium, does not undergo solid-state phase transformations so there are limited ways in which its microstructure can be controlled. One of these is to force marginally soluble solutes to form minute particles, i.e., to precipitate compounds. Copper is one of the suitable solutes because its solubility in aluminium is temperature sensitive (Figure 9.1). The precipitation reactions in Al-Cu are quite complex. The equilibrium phase Al_2Cu is difficult to nucleate so its formation is preceded by a series of metastable precipitates. The red lines on Figure 9.1 correspond to the metastable equilibrium between Al_{ss} and a variety of metastable precipitates including copper clusters, θ'' and θ'. The solubility of copper in the aluminium is greater when dealing with a metastable precipitate because the free energy of the metastable phase is greater than that of the equilibrium phase. Figure 9.2 illustrates this, with $Al_{ss}^{\theta} < Al_{ss}^{\theta'}$ as determined by the tangent construction.

An appropriate Al-Cu alloy quenched as a solid solution can start precipitating the copper in a very fine state at ambient temperature. The materials therefore hardens because the precipitates hinder plastic deformation. Some of these fine precipitates, which can only be resolved using atomic resolution, are shown in

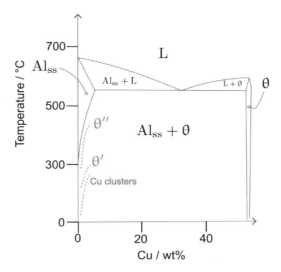

FIGURE 9.1 A part of the aluminium-copper phase diagram. The term Al_{ss} implies an aluminium-rich solution containing some copper, and $\theta \equiv Al_2Cu$. The dashed red lines represent the equilibrium between Al_{ss} on the left of each line and the copper-rich phase marked in red.

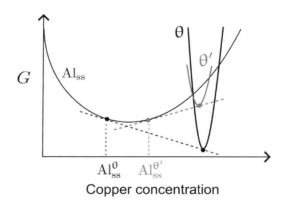

FIGURE 9.2 Free energy diagram showing why a metastable precipitate, which has a greater free energy than a stable precipitate, leads to an increased solubility of copper in the aluminium solid-solution. The compositions in equilibrium are, as usual, determined using the common tangent construction.

Figure 9.3.

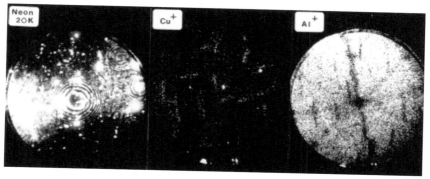

FIGURE 9.3 These images, which show individual atoms, were taken from Al-4Cu wt% solid solution that was aged at 130 °C for 95 h to produce the metastable transition zones illustrated by the Cu^+ image. The field ion image on the left shows all the atoms, the middle one only the Cu atoms and the one on the right, the Al atoms. Photographs courtesy of A. R. Waugh and S. Waugh. Many more relevant images of atoms in Al-Cu on https://www.phase-trans.msm.cam.ac.uk/2017/AlCu/AlCu.html

Example 26: why metastable precipitation?

Explain why metastable precipitates form at all, given that the formation of an equilibrium precipitate must lead to the greatest reduction of free energy.

Solution 26

Thermodynamics alone cannot explain why a less stable phase forms first. However, the equilibrium precipitate may take too long to form, in which case a metastable phase has the opportunity to form more rapidly, for example if it can nucleate much faster because it has a better lattice matching with the matrix – i.e., a lower interfacial energy per unit area. From Equation 6.2, the activation energy for nucleation is given by

$$G^* = \frac{16\pi\sigma_{\alpha\gamma}^3}{3(\Delta G_{CHEM} + \Delta G_{STRAIN})^2}$$

so G^* is more sensitive to the interfacial energy than to the driving force ΔG_{CHEM}, hence permitting a metastable phase to form first, albeit with a smaller reduction in free energy.

For example, iron is the most stable element in the universe [37], but other elements exist, until they eventually all decay towards iron. If you wish to see an interesting movie on the Al-Cu alloy *Duralumin* as it was used in airships during the first world war, see https://www.youtube.com/watch?v=TK9Q800EBAI (or search for Bhadeshia Cuffley Airship). It is a story for school children, explaining a metal fragment that was recovered from a crash site.

9.2 LIQUID CRYSTALS

Liquid crystals contain order that is associated with anisotropic molecules that can flow past each other and yet maintain a degree of orientational order. Normal liquids are isotropic. Figure 9.4a illustrates the difference between a crystal that has order in all directions, a liquid crystal which is imperfectly ordered and where the molecules can slide past each other along their long directions, and finally, a liquid that has no orientational order.

The molecules in a liquid crystal can often be aligned by an externally applied magnetic or electrical field. Transformation between the liquid and liquid-crystal states can also be induced by a change in temperature. A liquid crystal is optically active, one form of such activity is that it can bend the plane of polarisation of incident light to an angle corresponding to the long direction of the molecules. So by changing its orientation between crossed-polars, it is possible to construct a liquid-crystal display.

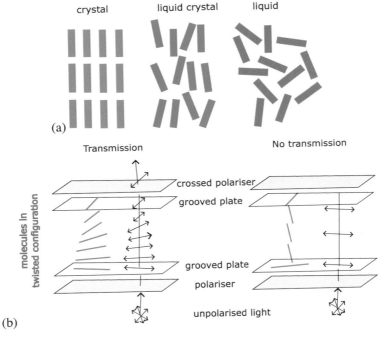

FIGURE 9.4 (a) A schematic illustration of the level of order associated with a liquid crystal. (b) An illustration of how a liquid crystal display works

Nematic liquid crystals are rather like the one displayed in Figure 9.4a, where the molecules are aligned approximately parallel but are not arranged in well-defined planes. If they were in well-defined planes, they would be called *smectic*. And there are other varieties. These phases can transform into one another when subjected to stimuli.

The working of a liquid crystal display is simply illustrated in Figure 9.4b. The grooves in the white plates are parallel to the plane of polarisation of the adjacent polarising membranes. The molecules initially lie parallel to the top and bottom grooved plates, and twist in-between. Polarised light entering from the bottom has its plane of polarisation twisted, finally parallel to the polariser at the top, thus allowing the transmission of light. In the diagram on the right, an electrical field is applied to change the orientations of the molecules which no longer twist the plane of polarisation of the incident light, thus not allowing transmission through the upper polariser. So the image switches from bright to dark. Such devices are used in watches, computer screens etc.

Example 27: cholesterol

When a particular, solid cholesterol-derivative (cholesteryl benzoate) is heated above 145 °C, it first forms a cloudy liquid and and then on reaching 170 °C, forms a clear liquid. And the transition from cloudy to clear is reversible. Explain why this might happen.

Solution 27

In fact, the cholesterol derivative from carrots was the first observation of a liquid crystal which forms as the solid turns into a liquid [38]. The liquid crystal has molecules aligned, but the alignment can change with position (rather like having small clusters of liquid crystals). This non-uniformity leads to the scattering of light, hence the cloudy appearance. As the temperature is raised, the molecules become disordered everywhere and therefore the liquid becomes clear since there are no 'domains' of ordered regions.

9.3 RANDOM QUESTIONS

1. A cube of gadolinium is sheared parallel to one of its faces, through an angle of 1°. Calculate the elastic energy per unit volume, stored within the material. (Young's modulus 55.5 GPa, shear modulus 22 GPa, density 7.9 Mg m^3).

2. The following data refer to the steady-state diffusion of carbon in a direction normal to a sheet of steel which is 0.2 mm thick. One of the surfaces is maintained at a constant concentration of carbon given by its equilibrium solubility with respect to the environment, and the other at zero concentration. Use the following data to derive the activation energy for diffusion.

Temperature / K	Solubility / g m^{-3}	Flux / g m^{-2} s^{-1}
1350	13000	3.67
1240	20000	2.5

3. In a pure substance, the chemical potentials of phases ϕ and ζ are given by:

$$\mu_\phi^\circ = 2140 - 20T \text{ J mol}^{-1}$$
$$\mu_\zeta^\circ = 4000 - 40T \text{ J mol}^{-1}$$

At what temperature are they in equilibrium? Which of these phases is stable at 100 K?

4. At the triple point on Figure 3.2 representing the equilibrium between ϵ, γ and α, the volume changes associated with the phases are given by [p.2, 39]

$$\left.\begin{array}{l} \Delta V(\alpha \to \epsilon) = -0.34 \\ \Delta V(\epsilon \to \gamma) = +0.13 \\ \Delta V(\gamma \to \alpha) = +0.21 \end{array}\right\} \quad \text{cm}^3 \text{ mol}^{-1}.$$

Which one of these solid phases of iron is expected at the centre of the planet Venus?

5. Prove that if

$$\left(\frac{\partial G}{\partial P}\right)_T = V$$

then

$$\left(\frac{\partial H}{\partial P}\right)_T = T\left(\frac{\partial S}{\partial P}\right)_T + V$$

6. For a crystalline sample, the Bragg equation predicts precisely the angles at which X-ray diffraction peaks in plots of intensity versus angle relative to the incident beam should be detected. Any deviation from the Bragg angle should yield zero intensity. Why then is the intensity associated with each peak spread out about the Bragg angle, resulting in broad diffraction peaks?

7. Figure 9.4a show three kinds of arrangements of anisotropic molecules. Rank these in terms of the breadth expected of any peak observed during an X-ray diffraction peak.

8. A transformation does not normally begin as soon as an equilibrium temperature is passed, because of the need to nucleate the new phase. A particular pure-substance can in principle undergo an $\Omega \rightarrow \beta$ transformation on cooling below 666 °C. The phase Ω and β are fully coherent. How much undercooling is needed to nucleate β before transformation proper can begin?

9. The figure below shows the free energy of mixing curves for two binary solutions at a particular temperature. Which one, given time, will become chemically heterogeneous? Will this be true at all temperatures?

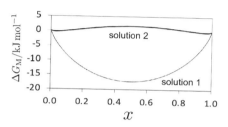

10. The diffusion coefficient of nitrogen in body-centred cubic-martensite with a lattice parameter 0.2866 nm is approximately $1.26 \times 10^{-7} \exp\{-73000/RT\}$ m^2 s^{-1}. Assuming that the nitrogen is located in octahedral interstices, calculate its maximum diffusion velocity at 500 °C.

11. During the cubic to tetragonal transformation in barium titanate, two of the oxygen atoms are displaced by -0.0085 nm, four oxygen atoms by -0.0065 nm and titanium by 0.006 nm. Calculate the resulting polarisation per unit volume, given that the volume of the unit cell is 0.0645 nm^3.

12. Large steel-reinforced concrete structures often are submerged in sea water; chloride ions within the sea water can penetrate the concrete by diffusion and attack the steel that is embedded within. If the salt concentration in the sea is 0.27% and the initial concentration within the concrete is zero, calculate the concentration expected at a depth of 1 cm following immersion for 6 months. The diffusivity of salt in concrete is 3×10^{-12} m^2 s^{-1}.

13. The cubic to tetragonal transformation in barium titanate occurs at the Curie point (p. 4). The respective lattice parameters are given by [40]:

$$a_{\text{cubic}} = 4.00692 + 0.1537 \times 10^{-4}T + 0.8277 \times 10^{-7}T^2$$

$$a_{\text{tetragonal}} = 3.99153 + 0.6284 \times 10^{-4}T - 0.3443 \times 10^{-7}T^2$$
$$+ 0.3495 \times 10^{-8}T^3$$

$$c_{\text{tetragonal}} = 4.03641 - 0.8647 \times 10^{-4}T + 0.9716 \times 10^{-6}T^2$$
$$- 0.9753 \times 10^{-8}T^3$$

where the parameters have units of Å and the temperature is in °C. Calculate the volume change during transformation at the Curie point. What would be the effect of hydrostatic pressure on the Curie point?

14. A substance undergoes a uniform expansion strain of 0.0006 while surrounded by a rigid container. If the bulk modulus is 100 GPa and Young's modulus 80 GPa, calculate the resulting elastic strain energy per unit volume.
15. Is the diffusion coefficient identical along all directions in a perfect crystal?
16. Given an activation energy G^* and a number per unit volume of nucleation sites N_V, derive an equation of the expected nucleation rate per unit volume per unit time (number per m^{-3} s^{-1}).

9.3.1 Brief answers to the big questions

1. $3.4 \times 10^6 \, \text{J m}^{-3}$.
2. The activation energy $Q = 102.87 \, \text{kJ mol}^{-1}$.
3. 93 K. The phase ζ is stable at 100 K.
4. ϵ.
5. Use $G = H - TS$.
6. If the lattice parameter of the crystal is not uniform, for example if the crystal is heterogeneously strained ... [41].
7. In order of increasing breadth, crystal, liquid crystal, liquid.
8. None.
9. Solution 2, no.
10. $0.01 \, \text{m s}^{-1}$.
11. $0.067e \, \text{C nm}$ where e is the charge on an electron (-1.602×10^{-19} C), divided by the volume of the unit cell, giving $0.165 \, \text{C m}^{-2}$.
12. $0.08 \, \%$.
13. $0.0407 \, \text{Å}^3$. The Curie point would decrease.
14. $18 \times 10^3 \, \text{J m}^{-3}$.
15. If the sites into which an atom can jump have cubic symmetry, the diffusion coefficient is constant. But in crystals of low symmetry, there may exist different varieties of sites and therefore the diffusion coefficient will vary with direction.
16. The nucleation rate per unit volume, $I_V \approx \nu N_V \exp\{-G^*/kT\}$, where ν is a attempt frequency. The Boltzmann constant is replaced by the gas constant depending on the units of G^*.

REFERENCES

1. E. Swallow, and H. K. D. H. Bhadeshia: 'High resolution observations of displacements caused by bainitic transformation', *Materials Science and Technology*, 1996, **12**, 121–125.
2. D. Meyerhofer: 'Transition to the ferroelectric state in barium titanate', *Physical Review*, 1958, **112**, 413–423.
3. B. Jaffe, W. Cook, and H. Jaffe: Piezoelectric Ceramics: Elsevier, 1971.
4. W. H. Bragg, and W. L. Bragg: 'The reflection of X-rays by crystals', *Proceedings fo the Royal Society A*, 1913, **88**, 428–438.
5. F. G. Caballero, and H. K. D. H. Bhadeshia: 'Very strong bainite', *Current Opinion in Solid State and Materials Science*, 2004, **8**, 251–257.
6. H. K. D. H. Bhadeshia: 'Nanostructured bainite', *Proceedings of the Royal Society of London A*, 2010, **466**, 3–18.
7. L. Kaufman, E. V. Clougherty, and R. J. Weiss: 'The lattice stability of metals – iii iron', *Acta Metallurgica*, 1963, **11**, 323–335.
8. W. Makower: 'XX. on a determination of the ratio of the specific heats at constant pressure and at constant volume for air and steam', *The London, Edinburgh, and Dublin Philosophical Magazine and Journal of Science*, 1903, **5**, 226–238.
9. W. Atwater, and J. Snell: 'Description of a bomb-calorimeter and method of its use.', *Journal of the American Chemical Society*, 1903, **25**, 659–699.
10. I. A. Bajenova, A. V. Guskov, P. G. Gagarin, A. V. Khvan, and K. S. Gavrichev: 'Experimental determination of the enthalpy of formation of the pyrochlore rare-earth hafnates', *Journal of the American Ceramic Society*, 2023, **106**, 3777–3791.
11. H. K. D. H. Bhadeshia: 'An introduction to thermal analysis techniques': https://www.phase-trans.msm.cam.ac.uk/2002/thermal.analysis.html, 2002.
12. Z. Mahbooba, L. Thorsson, M. Unosson, P. Skoglund, H. West, T. Horn, C. Rock, E. Vogli, and O. Harrysson: 'Additive manufacturing of an iron-based bulk metallic glass larger than the critical casting thickness', *Applied Materials Today*, 2018, **11**, 264–269.
13. S. Kriminski, C. Caylor, M. Nonato, K. Finkelstein, and R. Thorne: 'Flash-cooling and annealing of protein crystals', *Acta Crystallographica Section D: Biological Crystallography*, 2002, **58**, 459–471.
14. R. B. Schwarz, and W. L. Johnson: 'Remarks on solid state amorphizing transformations', *Journal of the Less Common Metals*, 1988, **140**, 1–6.
15. L. Boltzmann: 'Über die mechanische bedeutung des zweiten hauptsatzes der wärmegleichgewicht (on the mechanical meaning of the second law of thermal equilibrium)', *Wien. Ber*, 1877, **76**, 373–435.
16. L. Boltzmann: 'On some problems of the mechanical theory of heat', *Philosophical Magazine*, 1878, **5**, 236–237.
17. T. S. Chou, H. K. D. H. Bhadeshia, G. McColvin, and I. Elliott: 'Atomic structure of mechanically alloyed steels', In: *Structural applications of mechanical alloying*. Materials Park, Ohio, USA: ASM International, 1993:77–82.
18. J. S. Benjamin: 'Oxide dispersion strengthened ODS superalloys directional recrystallisation', *Metallurgical Transactions*, 1970, **1**, 2943–2951.
19. A. Y. Badmos, and H. K. D. H. Bhadeshia: 'Evolution of solutions', *Metallurgical & Materials Transactions A*, 2000, **28**, 2189–2193.
20. J. W. Yeh, S. K. Chen, S. J. Lin, J. Y. Gan, T. S. Chin, T. T. Shun, C. H. Tsau, and S. Y. Chang: 'Nanostructured high-entropy alloys with multiple principal elements: Novel alloy design concepts and outcomes', *Advanced Engineering Materials*, 2004, **6**, 299–303.

21. B. Cantor, I. T. H. Chang, P. Knight, and A. J. B. Vincent: 'Microstructural development in equiatomic multicomponent alloys', *Materials Science & Engineering A*, 2004, **375-377**, 213–218.

22. Z. Lou, P. Zhang, J. Zhu, L. Gong, J. Xu, Q. Chen, M. J. Reece, H. Yan, and F. Gao: 'A novel high-entropy perovskite ceramics $Sr_{0.9}La_{0.1}(Zr_{0.25}Sn_{0.25}Ti_{0.25}Hf_{0.25})O_3$ with low thermal conductivity and high Seebeck coefficient', *Journal of the European Ceramic Society*, 2022, **42**, 3480–3488.

23. L. Stixrude: 'Structure of iron to 1 Gbar and 40000 K', *Physical Review Letters*, 2012, **108**, 055505.

24. B. Cheng, S. Hamel, and M. Bethkenhagen: 'Thermodynamics of diamond formation from hydrocarbon mixtures in planets', *Nature Communications*, 2023, **14**, 1104.

25. B. Sundqvist: 'Carbon under pressure', *Physics Reports*, 2021, **909**, 1–73.

26. M. Ross: 'The ice layer in Uranus and Neptune – diamonds in the sky?', *Nature*, 1981, **292**, 435–436.

27. L. M. Ghiringhelli, J. H. Los, E. J. Meijer, A. Fasolino, and D. Frenkel: 'Modeling the phase diagram of carbon', *Physical review letters*, 2005, **94**, 145701.

28. T. W. Clyne: 'Microsegregation, cored dendrites': https://www.doitpoms.ac.uk/miclib/full_record.php?id=48, 2023.

29. C. E. Lundin, and M. J. Pool: 'Heats of mixing in praseodymium-neodymium solid solutions', *Journal of the Less-Common Metals*, 1965, **9**, 48–53.

30. J. H. Hildebrand: 'Solubility. xii. regular solutions', *Journal of the American Chemical Society*, 1929, **51**, 66–80.

31. E. A. Guggenheim: 'The statistical mechanics of regular solutions', *Proceedings of the Royal Society A*, 1935, **148**, 304–312.

32. Anonymous: 'Random walks': https://www.mit.edu/ kardar/teaching/projects/chemotaxis (AndreaSchmidt)/random.htm, 2023.

33. A. D. Smigelskas, and E. O. Kirkendall: 'Zinc diffusion in alpha brass', *Transactions of the AIME*, 1947, **171**, 130–142.

34. J. Rutter, and B. Chalmers: 'A prismatic substructure formed during solidification of metals', *Canadian Journal of Physics*, 1953, **31**, 15–39.

35. E. C. Bain: Alloying Elements in Steel: Cleveland, Ohio, USA, https://www.phase-trans.msm.cam.ac.uk/2004/Bain.Alloying/ecbain.html: American Society of Materials, 1939.

36. S. Chatterjee, and H. K. D. H. Bhadeshia: 'TRIP-assisted steels: cracking of high carbon martensite', *Materials Science and Technology*, 2006, **22**, 645–649.

37. E. M. Burbridge, G. R. Burbridge, W. A. Fowler, and F. Hoyle: 'Synthesis of the elements in stars', *Reviews of Modern Physics*, 1957, **29**, 547–650.

38. F. Reinitzer: 'Beiträge zur kenntniss des cholesterins', *Monatshefte für Chemie und verwandte Teile anderer Wissenschaften*, 1888, **9**, 421–441.

39. H. K. D. H. Bhadeshia: Theory of Transformations in Steels: London, U.K.: CRC Press, Taylor and Francis Group, 2021.

40. L. Shebanov: 'X-ray temperature study of crystallographic characteristics of barium titanate', *physica status solidi (a)*, 1981, **65**, 321–325.

41. B. D. Cullity, and S. R. Stock: Elements of X-ray diffraction: Third edition ed., Prentice-Hall, 2001.

Subject index

This short book is an introduction. Each of the concepts covered is sufficient and complete to the extent appropriate for a semester of an undergraduate course on phase changes in a wide variety of materials, from ferroelectrics to entropy ceramics and metals. Worked examples stimulate absorption of the subject. The book is suitable for second year students in any of the following disciplines: materials science, engineering, chemical engineering, physics, chemistry, Earth sciences and physical sciences in general. References are provided at the end for anyone who wants, for whatever reason, to delve deeper.

Harshad Bhadeshia and Haixue Yan are both academic staff committed to the creation and dissemination of knowledge at the School for Materials Science and Engineering, Queen Mary University of London

ISBN 9798395358547

90000

9 798395 358547

Printed in Great Britain
by Amazon

42943253R00069